AF387351

A. Yoshida (Ed.)

Smart Control of Turbulent Combustion

With 73 Figures

Springer

Springer Japan KK

Akira Yoshida, Ph.D.
Department of Mechanical Engineering, Tokyo Denki University
2-2 Kanda-Nishikicho, Chiyoda-ku, Tokyo 101-8457, Japan
e-mail: yoshida@yl.m.dendai.ac.jp

ISBN 978-4-431-66987-6   ISBN 978-4-431-66985-2 (eBook)
DOI 10.1007/ 978-4-431-66985-2

Library of Congress Cataloging-in-Publication Data

Smart control of turbulent combustion / A. Yoshida (ed.).
     p. cm.
  Proceedings of Workshop on Turbulent Premixed Combustion, held Sept. 14-15, 2000,
Nagoya Institute of Technology.
  Includes bibliographical references.

  1. Combustion engineering--Congresses. 2. Turbulence--Congresses. 3. Automatic
control--Congresses. I. Yoshida, A. (Akira), 1946- II. Workshop on Turbulent Premixed
Combustion (2000 : Nagoya Institute of Technology)

 TJ254.5 .S57 2001
 621.402'3--dc21

                                                         2001031430

Printed on acid-free paper

Typesetting: Camera-ready by the editors and authors
SPIN: 10795699

Preface

The International Workshop on Turbulent Combustion was held September 14–15, 2000, at the Nagoya Institute of Technology, to review the present status of turbulent combustion studies. Reviews were presented by Prof. F.A. Williams of the University of California, San Diego; Prof. Ken Bray of the University of Cambridge; and Prof. Jay Gore of Purdue University. Dr. Howard Baum of the National Institute of Standards and Technology and Dr. Jim McDonough of the University of Kentucky participated in the discussion. Some ten papers, describing the latest findings of Japanese studies in this field, were given at the meeting. About half of these studies are supported by a national project, the Open and Integrated Research Program, Creation of New Functionalized Thermo-Fluid Systems by Turbulence Control, that started only recently under the sponsorship of the Science and Technology Agency of Japan. The meeting was a great success and gave impetus and a sense of perspective to young Japanese researchers through the excellent reviews and valuable comments their work received.

I believe that this kind of open discussion is indispensable for any project to produce a good outcome, and I would like to extend my sincere thanks to all who participated in the meeting. Finally, I would like to express my special thanks to Prof. Tatsuya Hasegawa of the Nagoya Institute of Technology, Prof. Akira Yoshida of Tokyo Denki University, Prof. Kouichi Hayashi of Aoyama Gakuin University, and Dr. Satoru Ogawa of the National Aerospace Laboratory for their dedication to the success of the meeting.

I hope the publication of this volume, which collects most of the contributions at the workshop, will add to our understanding of turbulent combustion.

Tadao Takeno
Nagoya University

Contents

Some Recent Studies in Turbulent Combustion

F.A. Williams

Center for Energy Research, Department of Mechanical and Aerospace Engineering, University of California, San Diego, La Jolla, CA 92093-0411, U.S.A.

Summary. Turbulent combustion is complex. Many different approaches to describing turbulent combustion have been developed over the years. An attempt is made here to classify these approaches and to identify the regimes of turbulent combustion in which the different approaches may be most useful. Although the results are not entirely satisfactory, they may help to introduce some order into this highly complex topic.

Key words. Turbulent Combustion; Turbulence Modeling; Combustion Theory

Introduction

Turbulence and combustion are two complicated subjects. Combining them makes turbulent combustion exceedingly complex. The topic of this workshop thus may be expected to exhibit many different facets. This indeed is reflected in the range of papers in the present volume. They include both premixed and nonpremixed turbulent flames, experiment and theory, measurement techniques and results of measurements, computational results and theoretical interpretations – that is, they sample the spectrum of the topic. With such complexity, it is difficult to sort out the relationships among the various works.

A step towards bringing order into this chaotic complexity is to develop schemes for systematically classifying different approaches to the description of turbulent combustion. The present paper attempts to construct such a classification and to identify the kinds of turbulent combustion problems for which the various approaches may be best suited. A table of approaches is constructed, and a previously developed diagram of regimes of turbulent combustion is employed for the purpose of trying to identify where each approach may be most applicable and most useful.

Classification is difficult because many different approaches may be viewed in different ways. There is bound to be some controversy in assigning an approach to a specific category. Previous attempts at doing this by the author have been deemed by the author to be inadequate, and although the present classification appears to be an improvement, it still possesses deficiencies. It nevertheless may help to provide a kind of systematization that could aid in sorting the elements of this complicated topic. The classification therefore is presented here as a tentative scheme, to be considered in the future for further refinement and revision.

Approaches to turbulent combustion

Various current approaches to turbulent combustion can be found in books that have been published in recent years (Peters 2000, Libby and Williams 1994, Liñán and Williams 1993, Williams 1985[a], Williams 1985[b], Libby and Williams 1980). The six main headings in Table 1 represent one way to attempt to divide these approaches into different categories.

Phenomenological approaches are those that begin by considering turbulent combustion as a phenomenon in itself, independent of the underlying conservation equations, and then proceed to work various aspects of implications of the conservation equations into the phenomenological description.

Fluids-based approaches are those that, in a sense, begin in the opposite direction, starting with the underlying three-dimensional time-dependent conservation equations of Navier and Stokes with density variations and chemical reactions included (the reacting Navier-Stokes equations) and then, as necessary, work with various averages (moments) of these equations to obtain turbulence closure for describing turbulent combustion.

Perturbation-based approaches are those that employ perturbation theory in some sense or another, either perturbing the reacting Navier-Stokes equations directly or hypothesizing structures derived from perturbation theory and deriving possibly phenomenological conservation equations for those structures.

Imposition of the random process of turbulent statistics on the Navier-Stokes equations results in a linear and closed functional partial differential equation for the evolution of a probability-density functional (Hopf 1952), from which can be derived unclosed systems of partial differential equations for probability-density functions (PDFs); the same can be done for the reacting Navier-Stokes equations, and when sufficient modeling of the unclosed terms is introduced, closed PDF-evolution equations are obtained that can form the basis of descriptions of turbulent combustion, the fourth category in Table 1.

An alternative to working with such modeled evolution equations is to assume forms of the PDFs or to condition the PDFs in an effort to achieve better simplification prior to modeling, then employ moments from the second category for determining the presumed or conditioned PDFs, as listed in the fifth category.

Finally, there always seem to be methods that do not fall conveniently into any of these categories, and they are listed in the last category in Table 1, a category that is open-ended, since new types of approaches continually appear.

In the first category, zero-dimensional approaches (Mellor 1976, Heywood 1976), also called quasidimensional approaches, are among the oldest and, in many respects, most useful approaches; their utility stems from the fact that they are designed for specific applications, such as spark-ignition engines, but their large degree of empiricism prevents them from being extrapolated very well to conditions for which they were not initially adjusted. Age theories go beyond quasidimensional approaches by including times as a dimension; finding their main applications in chemical reactors, with residence times treated as random variables, they work mainly with ordinary differential equations (Pratt 1976). These theories have been extended further to include different zones in reactors separately or ultimately partial differential equations in space and time, as in Spalding's so-called ESCIMO approach (Spalding 1976, Ma et al. 1982). A new and in many ways very successful approach in the phenomenological category is the so-called one-dimensional turbulence of Kerstein (1999), derived from his linear-eddy concept (Kerstein 1988[a], 1990[b], 1991[c], 1992[d]).

The most direct of the fluids-based approaches, numerical integration (Jiménez et al. 1997, Boger et al. 1998) of the conservation equations (DNS), is limited in utility by computer capabilities. Currently a very active area of research is the extension of DNS to LES, where the necessary subgrid modeling is by far the principal source of difficulty (Germano et al. 1991, Ghosal andd Moin 1995, Girimaji and Zhou 1996, Cook 1997, Cook and Bushe 1999). Moment methods, the traditional approach for nonreacting turbulence, have been extended to reacting flows with closure at the first moments (algebraic closure) as well as at the second moments (Reynolds-stress closure). Most practical available computer programs, such as KIVA, Star CD, etc., adopt an intermediate approach, colloquially called closure at second order, where the only second-order moments considered are the turbulent kinetic energy k, the rate of dissipation of turbulent kinetic energy ε and the mean-square scalar fluctuation g. These various moment closures are widely used in describing turbulent combustion.

Perturbations-based approaches have employed a variety of different perturbation parameters. One such method is based on a dual expansion in turbulence intensity and scale for low-intensity, large scale turbulence (Clavin and Williams 1982, Aldredge and Williams 1991). Another (Peters 2000, Williams 1985[b], Markstein 1964, Clavin 1985) addresses premixed turbulent combustion in the limit in which the laminar flame thickness becomes small, the wrinkled-flame or flame-sheet limit, deriving an evolution equation for the flame sheet or reaction sheet, the so-called G equation, with the field variable G taking on a specified value a the reaction sheet, then working with moments of the random variable G to obtain a moment closure.[1] This same type of limit of thin reaction sheets has been taken as an explicit or implicit starting point for modeling of a more phenomenological nature, including the RIF approach for turbulent diffusion flames (Pitsch 1998) and the CFM approach for turbulent premixed flames (Candel and Poinsot 1990, Cant et al. 1990, Duclos et al. 1993).

In the category of use of PDF-evolution equations, one way to distinguish

4   F. A. Williams

approaches is on the basis of the kind of turbulent mixing model employed. Dopazo's linear mean-square estimation (Dopazo 1975), Curl's coalescence-dispersion (Curl 1963), Kraichnan's mapping closure (Chen et al. 1989) and Pope's Euclidiean minimum spanning tree (Subramanian and Pope 1998) are among the models adopted. Resulting predictions are similar in general but different in details, and each of the methods has certain deficiencies as well as certain favorable aspects to recommend it.

Presumed-PDF methods have been developed for the PDF of the mixture fraction Z for turbulent diffusion flames (Bilger 1976[a], Bilger 1976[b], Williams 1998) as well as, in premixed flames, for the PDF of a normalized reaction-progress variable c (the Bray-Moss-Libby, BML method (Bray and Moss 1977, Bray et al. 1984, Bray and Libby 1986, Bray et al. 1991)) and for the reaction-surface function G defined above (Peters 2000). Related ideas have recently been developed for partially premixed turbulent combustion (Peters 2000), combining elements of P(Z) and P(G), and there are similar generalizations with respect to P(Z) and P(c). Thin-flame or fast-reaction ideas underlie most of these approaches, as is evident in use of the random variable G. The conditional moment closer approach (Klimenko 1990, Klimenko and Bilger 1999, Bilger 2000) is another recent idea which comes from a somewhat different viewpoint but under suitable conditions can give results quite similar to those of the methods just mentioned.

Finally, there are approaches for premixed turbulent combustion that involve renormalization of the G equation (Yakhot 1988, Sivashinsky 1988) and appeal to the idea that in the wrinkled-flame limit the flame sheet may become a fractal (Gouldin 1987). In addition, such premixed turbulent flames have been treated as pseudosolitons (Tsugé to appear 2001). The continual appearance of new ideas suggests that Table 1 will have to be lengthened in the future.

Rejimes of turbulent combustion

Regimes of turbulent combustion can be viewed in different diagrams. Figure 1 is one such diagram that has been employed in recent years(Liñán and Williams 1993, Williams 1985, Libby abd Williams 1980, Williams 2000). It expresses the turbulence Reynolds number on the horizontal axis and an overall turbulence Damköhler number (the ratio of a turbulence time to a chemical time) on the vertical axis. For premixed turbulent combustion it is equivalent to the well-known Borghi diagram(Peters 2000) but with axes skewed. The different regimes indicated in the figure apply to premixed flames. The thin-reaction-zone regime, identified recently by Peters (2000), is included in the diagram. The coordinates selected in Fig. 1 have the advantage of being applicable to both premixed flames and diffusion flames, although specific attributes indicated in the figure pertain only to premixed flames. In general, however, irrespective of whether the system is premixed, nonpremixed or partially premixed, the weak-turbulence limit of wrinkled reaction sheets is at the top of the diagram and the strong-turbulent limit of distributed reactions a the bottom.

The diagram in Fig. 1 serves to illustrate that there are many different kinds of turbulent combustion depending on the regime, that is, on the location in the diagram.   Any one of the types of approaches discussed above may be expected to perform better in one of the regimes than in another. It is therefore of interest to try to identify where in the diagram each approach may best be applied. Sharp boundaries at edges of regions of applicability generally do not exist, but instead there is usually a gradual deterioration of any particular method in moving away from its optimum point of applicability.   Numbers have been placed in Fig. 1, corresponding to each of the approaches in Table 1, in the general vicinity of what is estimated to be the optimum point of applicability of the method. It is of interest to discuss the rationale for the placement of these numbers.

Category 1a applies best right in the center of a region of practical interest, but because of its empiricism it is good only over a narrow region in the diagram. The other entries in category 1 are seen to center below a Damköhler number of unity. This is because they are geared mostly to relatively slow chemistry, not addressing reaction sheets. They tend to move to larger Reynolds and Damköhler numbers in passing from 1b to 1c to 1d because of the additional elements that this sequence brings into the description. They are likely to have much broader ranges of applicability than methods in 1a, in particular extending to small Damköhler numbers, as a consequence of the character of their approximations. With suitable chemistry modeling, for example replacing real elementary chemistry by eddy-breakup types of ideas, some of these approaches also can be made to perform at higher Damköhler numbers.

Category 2a, being limited by what can be done with the computer, lies in the range of small Reynolds and Damköhler numbers and moves out of that basin only as computers evolve, not likely reaching the most practical regions of turbulent combustion. Extension to LES (2b) moves out and up, towards combustion reality, but there remain questions about subgrid modeling that may be expected to persist for a long time. In many respects, the physically most challenging problems are at the subgrid scale.   The methods in category 2 tend to apply over fairly broad regions about their optimum, and the optimums possibly increase in Reynolds number in moving from 2c to 2d to 2e, all of which seem to apply at least partly for Damköhler numbers greater than unity.

Almost by definition, method 3a holds at large Damköhler numbers and small Reynolds numbers; it certainly cannot reasonably be extended to Damköhler numbers approaching unity. Similarly, the other approaches in category 3 are for large Damköhler numbers, but these go to higher Reynolds numbers. The centers of 3c and 3d are placed somewhat nearer the centers of application because their modeling typically has been adjusted to put them there. The extents of the basins of applicability are, however, sufficiently broad that there is no well-justified reason for different placement of these methods.

Since all approaches in category 4 have about the same region of applicability, only one number 4 appears in Fig. 1. The character of the approach puts it at high Reynolds numbers and Damköhler numbers below unity. The breadth of the basin is, however, large enough to extend above a Damköhler number of unity. Many applications are, in fact, in the region of practical interest in turbulent combustion, and different approaches extend to different distances into this region. Special tricks, such as building in a reaction-sheet behavior, give results at high Damköhler numbers, but these are not general-purpose results.

Similar to category 4, there is only one number 5 in Fig. 1. It is located at about the same Reynolds number as category 4, perhaps a little lower because of the freedom of different assumptions concerning the PDF and the prevalent use of moment methods, but at a Damköhler number well above unity rather than below, giving much better correspondence with regions of interest in turbulent combustion. There is flexibility in that, for example for $P(Z)$, presumed shapes can be chosen that enable extension to significantly lower Damköhler numbers. This extension also applies to the CMC method, which in fact has been considered even for premixed combustion, although most of its applications pertain to nonpremixed systems, and which from a different viewpoint has elements in common with PDF-evolution methods which, as has been indicated, favor lower Damköhler numbers. Although $P(c)$ in principal also can be considered for development at lower Damköhler numbers, the $P(G)$ approach is strictly limited to the shaded area in Fig. 1, since G is not defined outside that region.

The first two of the entries in category 6 have been placed in the fractal region of Fig. 1 at the highest Reynolds numbers because this is where the reasoning that underlies them applies. They have been developed for premixed turbulent combustion, and they should not be expected to hold in either the single-sheet or thin-reaction-zone regime. The renormalization reasoning, right or wrong, at least implicitly involves fractal-like concepts, tying its applicability to that of fractals. Although fractal ideas have been discussed for nonpremixed turbulent combustion, they are at best multifractals in that situation, scalings different from those of ordinary fractals being needed; there are not yet any useable nonpremixed methods here. Entry 6c, unlike the other two in category 6, is not restricted to the fractal regime and in some respects is better suited to a single-sheet regime, where it has been placed in Fig. 1, although its specific regime is difficult to identify, and its basin of applicability could be very broad.

Conclusions and future prospects

Although an attempt has been made to be comprehensive in constructing Table 1, there surely exist approaches that have not been properly considered here. Moreover, additional approaches are sure to arise in the future. These new approaches can be placed under one of the six main headings in Table 1, quite obviously, since the category "other" is open to anything. It would be better to select one of the first five categories if possible, and since methods often have elements of more than one of these categories, placement can become a matter of judgment.

The various methods that currently exist span the entire plane of Fig. 1 in their regions of applicability; that is, no part of the plane has been overlooked in developing methods, and whatever the conditions of interest in an application may be, some potentially applicable method can be found for use there. Future work may be expected to be directed to improving descriptions in different regimes so that better results and better understanding can be obtained in specific regimes.

Trends in future work seem moving in two important directions. One is to retreat from considering only premixed and nonpremixed turbulent combustion by placing efforts on studying partially premixed systems (Peters 2000, Williams 2000), which certainly are of practical importance and which offer interesting challenges in modeling.

The other is to consider in one formulation different combustion processes that may exhibit both large and small Damköhler numbers simultaneously (Li et al. 1999). For example, in homogeneous-charge compression-ignition (HCCI) engines, as well as in spark-ignition engines, turbulent premixed flame propagation may occur at high Damköhler numbers while production of pollutants such as oxides of nitrogen occur behind this flame at low Damköhler numbers in the turbulent mixture. Methods for handling both of these phenomena and the interactions between them are currently being developed and improved, and additional notable progress may be anticipated, for HCCI engines as well as for other applications.

In general, turbulent combustion is a rich field of investigation, as the other papers in this volume demonstrate, with many exciting and new problems to be solved, both scientific and technological, advancing capabilities in energy utilization and conservation, propulsion and production of new materials, as well as fire and explosion safety.

Acknowledgment

This work is supported by the National Science Foundation through Grant No. NSF INT9815205.

References

Aldredge RC III, Williams FA (1991) Influence of wrinkled premixed-flame dynamics on large-scale, low-intensity turbulent flow. J Fluid Mech 228: 487-512

Bilger RW (1976)[a] The structure of diffusion flames. Combust Sci and Tech 13: 155-170

Bilger RW (1976)[b] Turbulent jet diffusion flames. Prog Energy Combust Sci 1: 87-109

Bilger RW (2000) Future progress in turbulent combustion research. Prog Energy Combust Sci 26: 367-380

Boger M, Veynante D, Boughanem H, Trouvé A (1998) Direct numerical simulation analysis of flame surface density concept for large eddy simulation of turbulent premixed combustion. Twenty-Seventh Symposium (International) on Combustion, The Combustion Institute, Pittsburgth, pp. 917-925

Bray KNC, Moss JB (1977) A unified statistical model of the premixed turbulent flame. Acta Astronautica 4: 291-319

Bray KNC, Libby PA, Moss JB (1984) Flamelet crossing frequencies and mean reaction rates in premixed turbulent combustion. Combust Sci and Tech 41: 143-172

Bray KNC, Libby PA (1986) Passage time and flamelet crossing frequencies in premixed turbulent combustion. Combust Sci and Tech 47: 253-274

Bray KNC, Champion M, Libby PA (1991) Premixed flames in stagnating turbulence; Part I. The general formulation for counterflowing streams and gradient models for turbulent transport. Combust Flame 84: 391-410

Candel S, Poinsot TJ (1990) Flame stretch and the balance equation for the flame area. Combust Sci and Tech 70: 1-15

Cant RS, Pope SB, Bray KNC (1990) Modelling of flamelet surface-to-volume ratio in turbulent premixed combustion. Twenty-Third Symposium (International) on Combustion, The Combustion Institute, Pittsburgh, pp. 809-815

Chen H, Chen S, Kraichnan RH (1989) Probability distribution for a stochastically advected scalar field. Phys Rev Lett 63: 2657-2660

Clavin P (1985) Dynamic behavior of premixed flame fronts in laminar and turbulent flows. Prog Energy Combust Sci 11: 1-59

Clavin P, Williams FA (1982) Effects of molecular diffusion and of thermal expansion on the structure and dynamics of premixed flames in turbulent flows of large scale and low intensity. J Fluid Mech 116: 251-282

Cook AW (1997) Determination of the constant coefficient in scale similarity models of turbulence. Phys Fluids 9: 1485-1487

Cook AW, Bushe WK (1999) A subgrid-scale model for the scalar dissipation rate in nonpremixed combustion. Phys Fluids 11, 746-748

Curl RL (1963) Dispersed phase mixing: I Theory and effects in single reactors. AICHE J 9: 175-181

Dopazo C (1975) Probability density function approach for a turbulent axisymmetric heated jet. Centerline evolution. *Phys Fluids* 18: 397-404

Duclos JM, Veynante D, Poinsot TJ (1993) A comparison of flamelet models for premixed turbulent combustion. Combust Flame 95: 101-117

Germano M Piomelli U Moin P, Cabot WH (1991) A dynamic subgrid scale eddy viscosity model. Phys Fluids A3: 1760-1765

Girimaji SS, Zhou Y (1996) Analysis and modeling of subgrid scalar mixing using numerical data. Phys Fluids A8: 1224-1236

Ghosal S, Moin P (1995) The basic equations for the large eddy simulation of turbulent flows in complex geometry. J Comput Phys 118: 24-37

Gouldin FC (1987) An application of fractals to modeling premixed turbulent flames. Combust Flame 68: 249-266

Heywood JB (1976) Pollutant formation and control in spark-ignition engines. Prog Energy Combust Sci 1: 135-164

Hopf E (1952) Statistical Hydromechanics and Functional Calculus. J Rat Mech Anal 1: 87-123

Jiménez J, Liñán A, Rogers MM and Higuera FJ (1997) A priori testing of subgrid models for chemically reacting non-premixed turbulent shear flows. J Fluid Mech 349: 149-171

Kerstein AR (1999) One-dimensional turbulence: Model formulation and application to homogeneous turbulence, shear flows and buoyant statistical flows. J Fluid Mech 392: 277-334

Kerstein AR (1988)[a] Linear-eddy model of turbulent transport and mixing. Combust Sci and Tech 60: 391-421

Kerstein AR (1990)[b] Linear-eddy modelling of turbulent transport. Part 3: Mixing and differential molecular diffusion in round jets. J Fluid Mech 216: 411-435

Kerstein AR (1991)[c] Linear-eddy modelling of turbulent transport, part 6. Microstructure of diffusive scalar mixing fields. J Fluid Mech 231: 361-394

Kerstein AR (1992)[d] Linear-eddy modelling of turbulent transport. Part 7. Finite-rate chemistry and multi-stream mixing. J Fluid Mech 240: 289-313

Klimenko AY (1990) Multicomponent diffusion for various scalars in turbulent flows. Fluid Dyn 25: 327-334

Klimenko AY, Bilger RW (1999) Conditional moment closure for turbulent combustion. Prog Energy Combust Sci 25: 595-687

Li SC, Williams FA, Gebert K (1999) A simplified, fundamentally based method for calculating NOx emissions in lean premixed combustors. Combust Flame 119: 367-373

Libby PA, Williams FA (1980) Turbulent Reacting Flows, Springer-Verlag, Berlin

Libby PA, Williams FA (1994) Turbulent Reacting Flows. Academic Press, London

Liñán A, Williams FA (1993) Fundamental Aspects of Combustion. Oxford University Press, New York, pp. 111-151.

Ma ASC, Spalding DB, Sun RLT (1982) Application of ESCIMO to the turbulent hydrogen-air diffusion flame. Nineteenth symposium (International) on Combustion, The Combustion Institute, Pittsburgh, pp. 393-402

Markstein GH (1964) Non-Steady Flame Propagation. Macmillan, New York, p.8

Mellor AM (1976) Gas turbine engine combustion. Prog Energy Combust Sci 1: 111-133

Peters N (2000) Turbulent Combustion. Cambridge University Press, Cambridge

Pitsch H (1998) Modellierung der Zündung und Schadstoffbildung bei der dieselmotorischen Verbrennung mit Hilfe eines interaktiven Flamelet-Modells. Dissertation, RWTH Aachen

Pratt DT (1976) Mixing and chemical reaction in continuous combustion. Prog Energy Combust Sci 1: 73-86

Sivashinsky GI (1988) Cascade-renormalization theory of turbulent flame speed. Combust Sci and Tech 62: 77-96

Spalding DB (1976) Mathematical models of turbulent flames: A review. Combust Sci and Tech 13: 3-25

Subramaniam S, Pope SB (1998) A mixing model for turbulent reactive flows based on Euclidian minimum spanning trees. Combust Flame 115: 487-514

Tsugé S (to appear, 2001) Rate increase in chemical reaction and its variance under turbulent equilibrium. Combust Sci and Tech

Williams FA (1985)[a] Combustion Theory. Addison-Wesley, Redwood City, CA, pp. 373-445

Williams FA (1985)[b] Turbulent combustion, The Mathematics of Combustion. J.D. Buckmaster, editor, SIAM, Philadelphia

Williams FA (1998) Formulation of turbulent diffusion flames with reduced chemistry. Combust Sci and Tech Japanese edition Supplement 6: 37-46

Williams FA (2000) Progress in knowledge of flamelet structure and extinction. Prog Energy Combust Sci 26: 657-682

Yakhot V (1988) Propagation velocity of premixed turbulent flames. Combust Sci and Tech 60, 191-214

Table 1: Approaches to Turbulent Combustion

1. Phenomenological
    a. Zero-Dimensional; Quasidimensional
    b. Age Theories; Chemical Reactors
    c. Multidimensional Age Theories; ESCIMO
    d. Linear Eddy; One-Dimensional Turbulence
2. Fluids-Based
    a. Direct Numerical Simulation (DNS)
    b. Large-Eddy Simulation (LES)
    c. Algebraic Closure (1'st Order)
    d. k-ε(-g) Modeling (1 1/2 Order) KIVA, Star CD, SPIDER, FIRE, SPEED, ...
    e. Reynolds-Stress Closure (2'nd Order)
3. Perturbution-Based
    a. Low-Intensity, Large-Scale Pertubation
    b. G-Equation Moment Modeling
    c. Flamelet Libraries (RIF) (Representative Interactive Flamelets)
    d. Flame-Surface Evolution Modeling (CFM) (Coherent Flamelet Model)
4. PDF Evolution
    a. Linear Mean-Square Estimation (Interaction by Exchange with Mean)
    b. Coalescece-Dispersion
    c. Mapping Closure
    d. Euclidean Minimum Spanning Tree
5. Presumed or Conditioned PDF
    a. P(Z) for Diffusion Flames
    b. P(c) for Premixed Flames
    c. P(G) for Premixed Flames
    d. Conditional Moment Closure (CMC)
6. Other
    a. G-Equation Renormalization
    b. Fractals
    c. Pseudosolitons

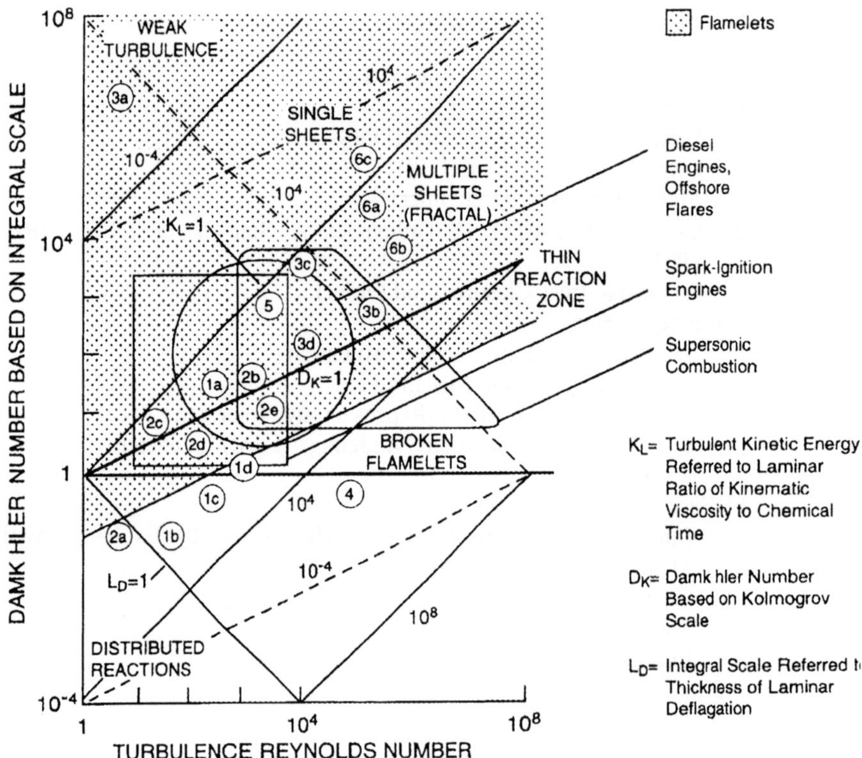

Figure 1. General regimes of turbulent combustion showing where different approaches of Table 1 may best apply.

Local Chemiluminescence Measurements of OH*, CH* and C$_2$* at Turbulent Premixed Flame-Fronts

Yuji Ikeda, Jun Kojima, and Tsuyoshi Nakajima

Department of Mechanical Engineering, Kobe University, 1-1 Rokko, Nada, Kobe 657-8501, Japan

Summary. Local chemiluminescence measurements of OH*, CH* and C$_2$* were carried out at the flame-front of a premixed turbulent methane flame to understand the details of the reaction zone and flame-front structures. Cassegrain optics were used to detect these chemiluminescences simultaneously. The developed system allows us to measure each chemiluminescence at high temporal resolution of 4 μs and high spatial resolution of 0.1 mm in diameter and 0.8 mm long. The relationship between the turbulent scale and the flame-front thickness was investigated by simultaneous measurements of these three chemiluminescences as well as those derived from LDV. It was found that the time series chemiluminescences could describe the flame-front duration and inter-arrival time, and that the variations in OH*, CH* and C$_2$* chemiluminescences occurred simultaneously, which indicates the strong relationship between OH* to CH* and OH* to C$_2$* reactions at the flame-front. Furthermore, the flame-front structure could be characterised by the time scales of chemiluminescence intensities, which was found to correspond to the flame-front thickness and the local Damköhler number. A new measurement technique to measure the local flame-front structure by means of local chemiluminescence measurements was proposed and its performance was proven.

Key words. Turbulent Premixed Flames, Local Flame-Front Structure, Chemiluminescence, Cassegrain Optics, Local Damköhler Number

Introduction

Numerous studies have been made regarding the effects of turbulent flow and its properties on flame-front structure and chemical reactions (Damköhler et al. 1940, Summerfield et al. 1955, Karlovitz et al. 1951, Ballal 1979). Turbulent premixed flames may be defined by their characteristic local flame-front structure and the combustion reaction rate at this flame-front (also referred to as the "reaction zone") (Williams 1985). The flame-front travels from burned to unburned gas with a shape and size that is primarily dictated by turbulence, but it is also affected by characteristics of the chemical reaction such as the heat release rate, temperature gradient, reaction rate, preferential diffusion, radiation, and so on. Relationships to turbulent scales such as the Kolomogorov scale and reaction zone thickness (Furukawa et al. 1990, Bedat et al. 1995) have been widely used to classify the features of turbulent flame structures. There are several theories regarding these flames. The first is that the flame-front structure is similar to that of a wrinkled laminar flame when the turbulent intensity is weak and its scale is smaller than the flame-front thickness (Buschman et al. 1996). Another theory is that the turbulent flame consists of the distributed reaction zone when the turbulent intensity is extremely high (Summerfield et al. 1955, Yoshida et al. 1992).

The flame-front thickness has been reported by several researchers as less than 1 mm (Furukawa et al. 1990, Buschmann et al. 1996). The flame-front may be defined by the several physical or chemical quantities present in this area.

Many researchers have emphasized the importance of precise and detailed experimental data for the flame-front structure and have tried to obtain this data, but it has not been measured directly. If possible, this data would determine the flame structure in space and its evolution in time.

The flame-front is not fixed in space, but rather travels through the turbulent flow field of unburned fuel. The turbulent scale is related to the flame-front movement, but not directly. The movement and shape of the flame-front do not coincide with the turbulent eddies. They are instead determined by reaction speed, turbulent flow characteristics, thermo-chemical properties, and heat balance.

Local flame-front structures have been investigated to determine the effects of turbulence-chemistry interaction (Kuznetsov et al. 1990), vortex interaction with the surface (Tabaczynski et al. 1978), stretch rate (Kuo 1986), and so on. Analysis of this data has almost always been performed using time-averaged statistics.

Laser techniques have greatly enhanced researchers' ability to measure the flame-front structure and its features. Laser-induced fluorescence (LIF) (Hanson et al. 1988, Deschamps et al. 1996) has been a very powerful tool to visualize the front of OH or CH concentration and its evolution over time (Nguyen et al. 1996, Renfro et al. 2000). Laser Doppler Velocimeter (LDV) measurements have been used to identify the turbulent scale and its intensity. This is made possible by the instrument's fine spatial resolution of 50 μm. Thermocouple or electrostatic probe have been used to characterize the effect of turbulence on the flame configuration or burning velocity (Yoshida et al. 1982, Furukawa 1998). Unfortunately, intrusive probes may disturb the flow field. In addition, it is impossible to measure the same point in space with the LDV, thermocouple, and electrostatic probe at the same time.

Current measurement techniques can provide either detailed information on the instantaneous spatial structure of the flame with high temporal resolution (LIF, Schlieren, etc.), or temporal behavior with fine spatial resolution for a single point in space (LDV, thermocouple, electrostatic probe, etc.).   It is clear that compromises must be made in order to study the temporal and spatial structure of turbulent flames.

The chemiluminescence of electronically excited radicals in a flame results from chemical reactions (Gaydon 1974). It is known that the emission intensity from OH*, CH*, and C_2* (here the * denotes an electronically excited state) in hydrocarbon flames can be used to observe the location of the primary combustion region in time-series with relatively easier and compact measurement system. Recent experimental and numerical studies have investigated several meaningful quantitative correlations between chemiluminescent emission intensities and flame structural parameters such as species concentration, temperature or equivalence ratio (Samaniego et al. 1995, Roby et al. 1998, Docquier et al. 2000). But most of these chemiluminescence measurements collect the global emissions along the line-of-sight of the typical collecting lens optics and there is not enough spatial resolution to detect the local flame front.   There has been little study of the absolute concentrations of these species and its spatially resolved structure (Walsh et al. 1998, Najm et al. 1998, Samaniego et al. 1999).   Such measurements of chemiluminescence are needed in reaction modeling to predict excited radical concentrations and in understanding the structure of the reaction zone.

We developed an optical measurement technique using specially designed Cassegrain Optics to measure the spatially resolved flame emissions (Akamatsu et al. 1999). This Cassegrain mirror systems have been used for flame emission measurements in spray flames, internal combustion engines, and so on (Ikeda et al. 1998, Tsushima et al. 1998). In our previous study, we have measured OH*, CH* and C_2* chemiluminescence at the reaction zone of laminar premixed flames (Kojima et al. 2000) and also turbulent propane/air premixed flames (Kojima et al. 1999, Ikeda et al. 2000). We have to demonstrate these techniques can apply to other flames with different kind of fuel and conditions.

The purpose of this research is to examine the local flame-front structure of turbulent premixed flames using spatially and temporally resolved measurement of chemiluminescence with LDV technique. The simultaneous measurements of OH*, CH* and C_2* chemiluminescence were carried out in a Bunsen burner (CH_4/air, Inner diameter = 20 mm, Re = 2400, ϕ = 0.95, u'/S_L = 0.4).

Experimental apparatus

The experiments were carried out using a turbulent premixed burner of 20 mm inner diameter (R = 10 mm), Reynolds number (Re) of 2400, mean velocity (U) of 2 m/s, and an turbulent intensity (u') of 0.15 m/s. Methane was used as the fuel at an equivalence ratio of 0.95. u'/S_L is 0.4, where S_L is laminar burning velocity. Figure 1 shows a laser tomography image of the flame created by a Cu-vapor laser and a synchronized high speed camera (1/4500 fps).

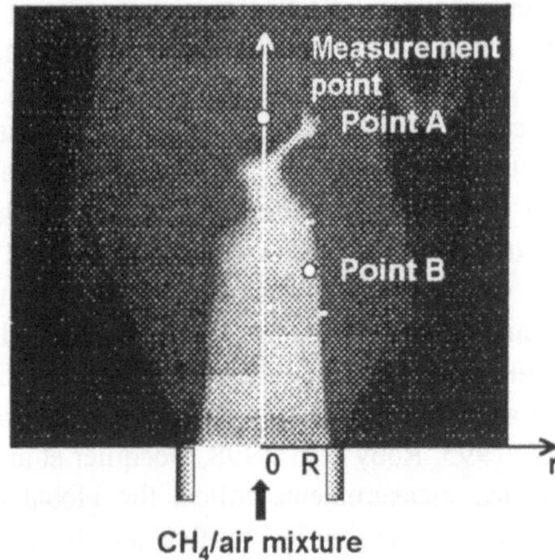

Fig. 1. Laser tomography image of the turbulent premixed flames (Burner: CH$_4$/air, Re = 2400, equivalence ratio ϕ = 0.95, nozzle diameter 20 mm. CCD camera: 1/4500 fps)

The Cassegrain optics were developed (Akamatsu et al. 1999) and their measurement volume dimensions were calculated and measured as shown in Fig. 2. A small measurement volume of ϕ 0.1 x 0.8 mm was achieved to produce high spatial resolution similar to that of the LDV. The simultaneous measurement system for local chemiluminescences of OH*, CH* and C$_2$* and LDV is shown in Fig. 3. Chemiluminescence emission received by the Cassegrain optics was focused onto an optical quartz fiber (core: 0.2 mm diameter) and transmitted to a spectroscopy unit. As shown in the figure, color splitters (interference and dicroic filters) for OH*, CH*, C$_2$* were implemented for each species. The specifications for each of the optical filters was as follows (center wavelength / half-band width / transmitting efficiency):

OH*: 306nm / 14nm / 61%
CH*: 431.4nm / 1.5nm / 40%
C$_2$*: 516.5nm / 2nm / 58%

The time-series intensity signals of the three chemiluminescences were measured across the flame-front at the "flame-cone" (x/D = 1.5, r/R = 0.8) and at the "flame-tip" (x/D = 3.0, r/R = 0).

The chemiluminescence was fed through the three optical filters and terminated in three photo-multipliers (Hamamatsu R212U). The analog output signals of these receivers were amplified, filtered, and digitized by a 250 kHz A/D converter. The large memory capacity of the system permitted time-series measurements over several seconds.

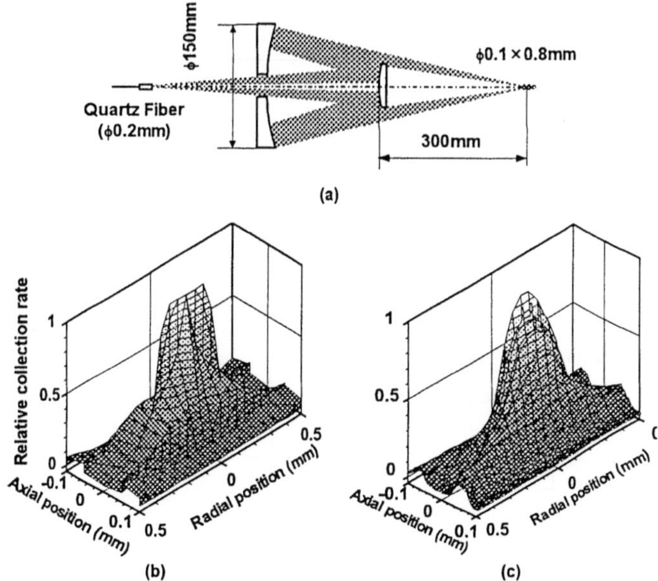

Fig. 2a,b,c. Developed Cassegrain optics. **a** Cassegrain system consists of a concave mirror (ϕ = 150 mm), convex mirror (ϕ = 50 mm) and quartz fiber (core diameter 0.2 mm). **b** Calculated collection rate distribution by Akamatsu et al. 1999. **c** Measured collection rate distribution by Kojima et al. 1999.

Results and discussion

The spectroscopy data of this turbulent premixed flame were measured by spectrometer (Mcpherson, 0.35 Meter Monochrometer, 150 G/mm) with CCD (Princeton, ICCD-576G) as shown in Fig. 4 (a). Figure 4 (b) shows the measured local flame spectra of laminar flame-front in our previous study by Kojima et al. (2000). These plots show the three typical bands corresponding to OH*, CH*, and C_2* emission. We have demonstrated the emission intensity ratio of OH*/CH*, C_2*/CH* and C_2*/OH* could be a good marker for local mixture strength at the flame-front over a range of ϕ = 0.9 to 1.5.

The typical time-series signals of local chemiluminescence intensities and axial velocity at the flame tip are shown in Fig. 5. It is clear that a band corresponding to OH* exists along with those for CH* and C_2*. A significant coincidence was observed between emissions from OH*, CH*, and C_2*. The characteristics of the time variation and intermittence of these three signals are very much

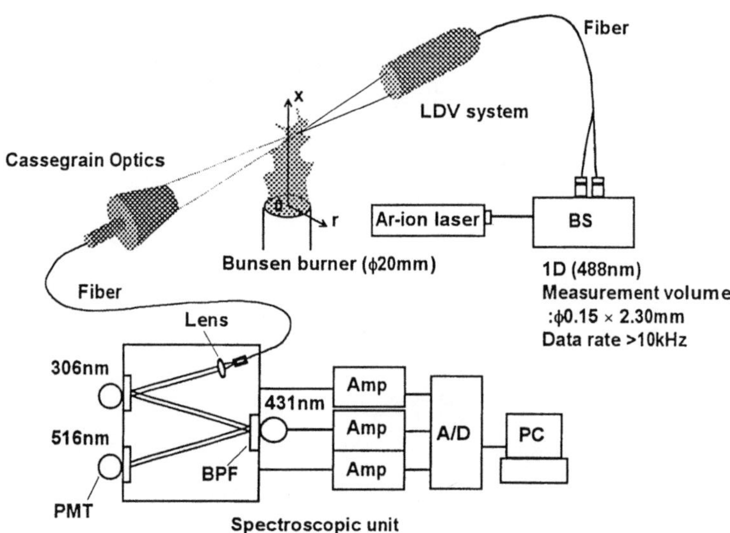

Fig. 3. Time-series measurement of local OH*, CH* and C_2* chemiluminescece intensities by Cassegrain spectroscopic system and velocity by LDV.

synchronized. The peaks in these chemiluminescences were not of the same amplitude, but their flame-front durations can be easily obtained. The flame-front durations and emission intensities are related to their reaction rates. The flame-front duration of each species can be obtained by measuring a signal's residency time above a certain threshold level. This time scale can be converted to the length scale of flame-front thickness with the mean velocity at the point. When OH* emission was produced, CH* and C_2* were also generated. Using a certain threshold level, these three signals were digitized to calculate the duration of the OH* signal and its inter-arrival time.

Local flame-front movement can be measured by examining the relationships of velocity gradients and emission intensity peaks as shown in Fig. 5. Remarkable velocity changes over 2 m/s are observed when the flame-front passes through the measurement point. From these values we can estimate the local axial velocity just behind and beyond the flame-front are about 0.9 m/s and 3.0 m/s. Furthermore, the unburned gas side or the burned gas side of the instantaneous local reaction zone passing the measurement point can be determined by this data.

OH* chemiluminescence intensities were measured across the flame-front in radical direction from x/D = 0.2 to 1.5 and their probability density functions (PDF) were calculated (Fig. 6). In the averaged flame zone, the PDF shows a bi-modal profile at r/R = 0.8 and 0.9. The PDFs at inner and outer locations (r/R < 0.6 and r/R = 1.0) does not show bi-modal peaks. This kind of bi-modal peak was

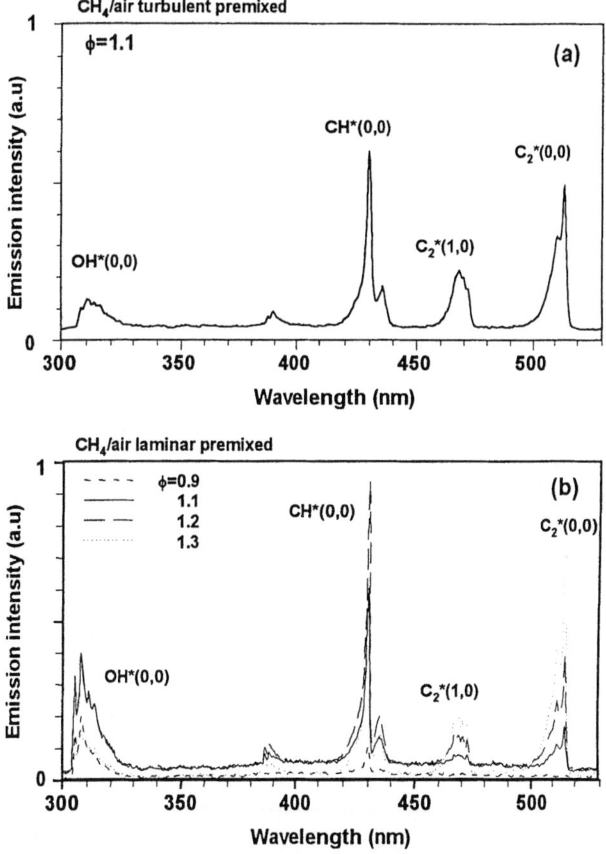

Fig. 4a,b. Local flame spectra measured at the flame-front. a Methane/air turbulent premixed flames. b Methane/air laminar premixed flames at the different equivalence ratio.

also noted in temperature measurements (Yoshida et al. 1982) and it was noted that these profiles were typical of wrinkled laminar flames. Since the turbulent intensity of this burner was relatively low (Re = 2400), the flame-front may display the characteristics of a wrinkled laminar flame. The local chemiluminescence data was obtained non-intrusively with a high temporal resolution of 4 µs, thereby permitting the co-existence of small measurement volumes and small time steps.

Flame-front thickness will be used here as an indication of flame-front structure. Figure 7 shows the calculation of these values. The chemiluminescence data was converted into a digital signal reflecting whether the signal was above or below a threshold value. The duration of these pulses t_f and their inter-arrival

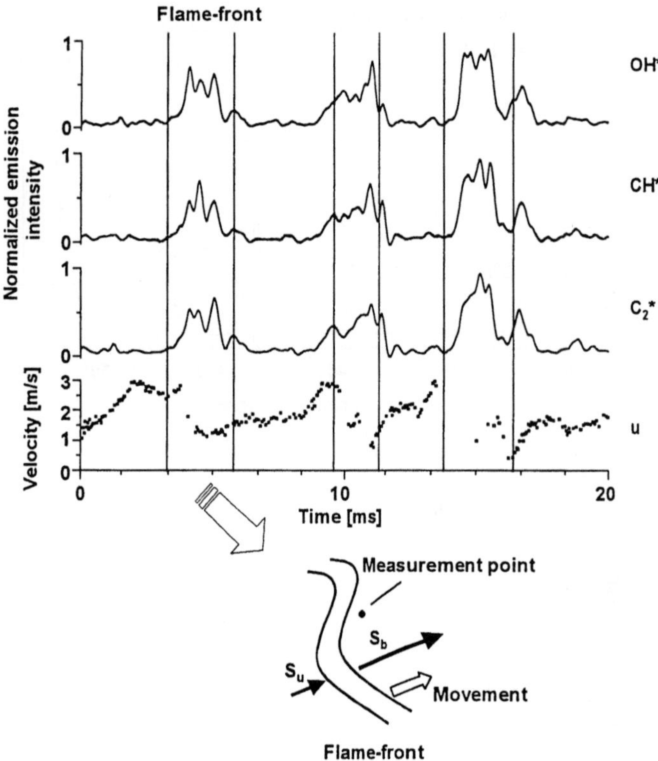

Fig. 5. Typical time-series signals of local OH*, CH* and C_2* chemiluminescence and axial velocity at the turbulent flame-front ($r/R = 0.8$). Mean velocity U = 2.0 m/s, turbulent intensity u' = 0.15 m/s. Here, S_u means unburned gas velocity and S_b means burned gas velocity.

times t_{int} were calculated and histograms of these values were plotted. Mean values for t_f and t_{int} were calculated from these histograms and these were multiplied by mean axial velocity from the LDV to calculate flame-front thickness δ_L and its scale.

As shown in Fig. 5, velocity measurement at the same point of chemiluminescence was done. The velocity histograms in approach flow indicate that flow is turbulent as shown in Fig. 8. The power spectrum of the approach flow shows significant profile of –3/5 power law. And the velocity histogram at the flame front demonstrate bi-modal velocity peaks, which can support to understand the flame is laminar flamelet as well as these found in Fig. 6.

The flame-front thickness calculated from the three chemiluminescences are shown in Fig. 9 with histograms for two locations. The flame-front thickness δ_L

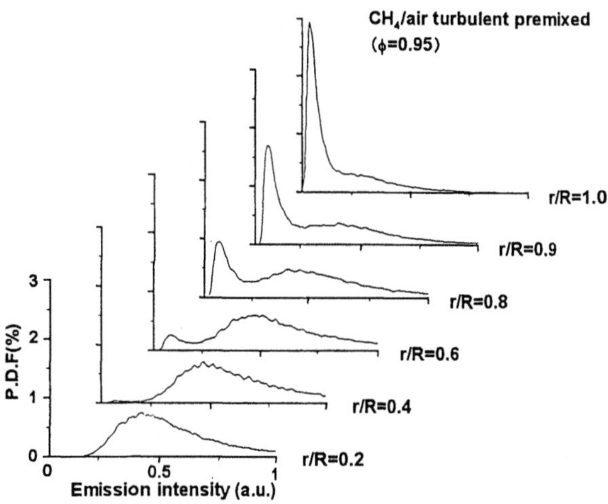

Fig. 6. Probability density functions (PDFs) of local OH* chemiluminescence intensity at the different location along with radial direction.

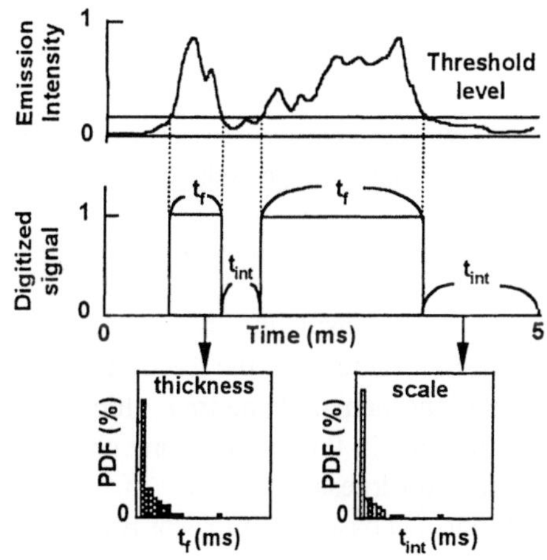

Fig. 7. Scheme to calculate flame-front thickness and its scale from time-series signal

calculated at two points using the OH* data was 2.2 and 2.4 mm, but the thickness by using CH* and C$_2$* data were almost one-tenth this size (0.24 to 0.31mm). The results of using the CH* and C$_2$* data are similar to those reported by Buschmann et al (1996). The discrepancy between the results calculated using the different

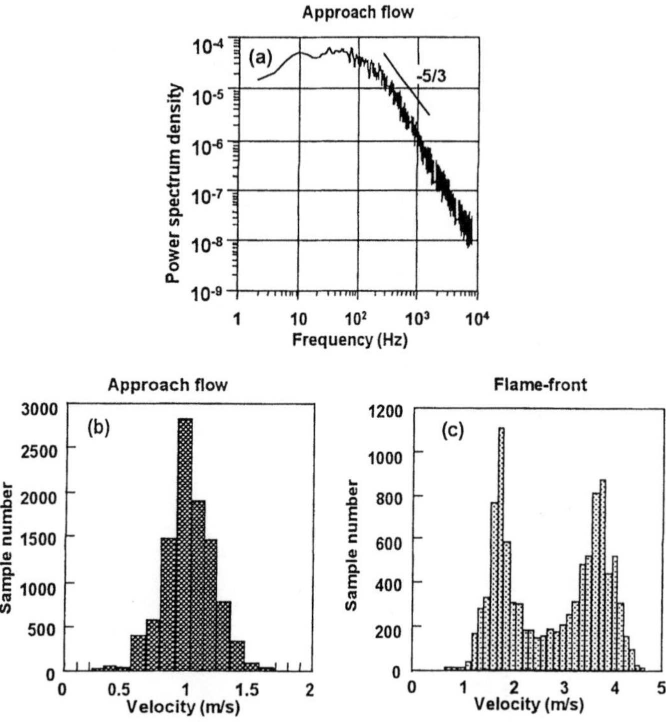

Fig. 8a,b,c. Turbulent flow characteristics. **a** Power spectrum density of unburned approach flow. **b** Histogram of unburned gas velocity at the burner port. **c** Histogram of gas velocity at the flame-front.

species can be understood by considering that the OH* chemiluminescence can exist in the high temperature region and the reaction zone, so that its distribution area may be wider than the actual flame-front thickness. The CH* and C_2* chemiluminescences exist only in the reaction zone so their durations can be used to measure the flame-front thickness.

The Damköhler number, *Da* based on turbulent micro scale was also calculated and shown in Table 1. *Da* between 3.6 and 4.7 were found using CH* and C_2* data, while those derived from OH* data were an order of magnitude lower due to unexpected wider thickness of OH* front. *Da* of CH* and C_2* are almost the same as that measured in our previous study on propane/air premixed turbulent flames (Ikeda et al. 2000). Using the same procedure for inter-arrival times, the scale of the flame-front was calculated as shown in Table 1. The flame-front thickness at the flame tip was almost ten-times larger than that at $x/D = 1.5$.

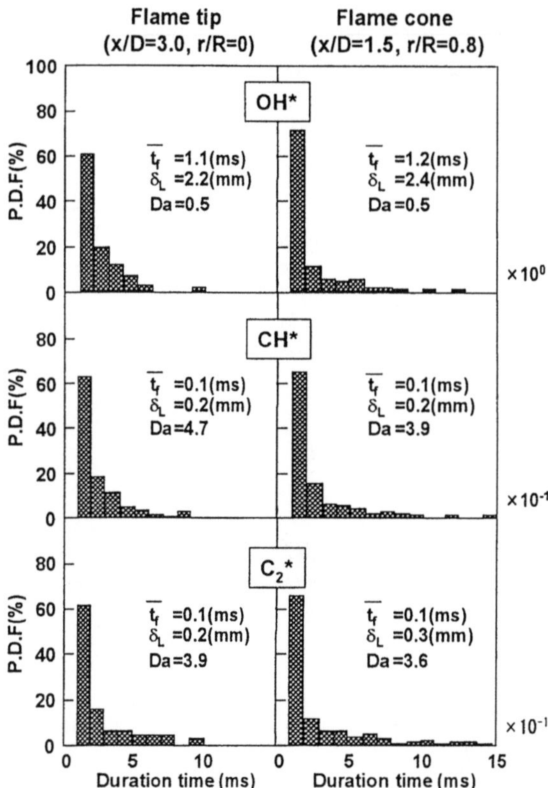

Fig. 9. Histograms of instantaneous local flame thickness of OH*, CH* and C_2* front estimated by the duration time of chemiluminescence intensity signals.

To help in understanding the time scale of flame-front thickness and temporal behavior, auto-correlation functions of the three chemiluminescences were generated as shown in Fig. 10. The integral time scale of the OH* data was larger than those of the CH* and C_2* results. The time scales of the CH* and C_2* species at different locations were nearly identical. This result further demonstrates that the flame-front thickness and its scale should be examined through the use of CH* and C_2* chemiluminescences, rather than those of OH*. Taylor micro time scale of turbulence measured by the LDV was 0.23 ms at the flame tip, which corresponds to an length scale, λ_t of 0.46 mm. It was found that micro length scale was larger than the reaction zone thickness.

Table 1. Flame-front thickness, its scale and local Damköhler number based on OH*, CH*, C_2* chemiluminescence. Here, Damköhler number is defined as $Da = \lambda_t \cdot S_L / u' \cdot \delta_L$, where δ_L is flame thickness (mm), S_L is laminar burning velocity (m/s), λ_t is turbulent Taylor micro scale (mm), u' is turbulent intensity (m/s).

Structuaral parameter	Location	OH*	CH*	C_2*
Flame-front thickness (mm)	Tip	2.2	0.24	0.30
	Cone	2.2	0.30	0.31
Flame-front scale (mm)	Tip	4.3	4.9	4.8
	Cone	0.61	1.1	1.8
Damköhler number	Tip	0.50	4.7	3.9
	Cone	0.47	3.9	3.6

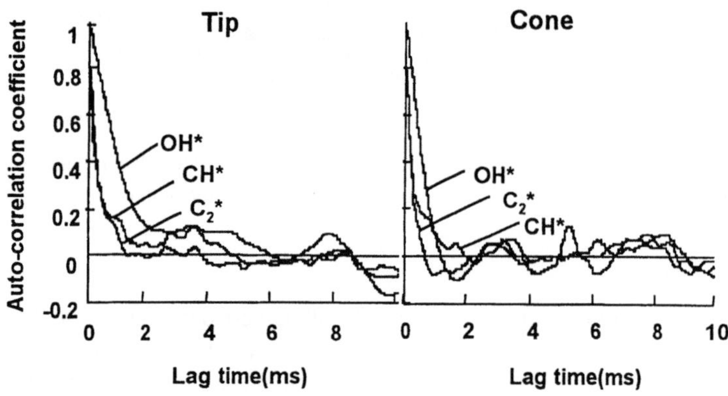

Fig. 10. Auto-correlation coefficients of time variations of OH*, CH* and C_2* chemiluminescence intensities.

The relationship between the flame-front thickness and the turbulent scale was investigated. For weak turbulence in the present study, the turbulent scale is big in comparison to the laminar flame thickness. Strictly speaking, the question

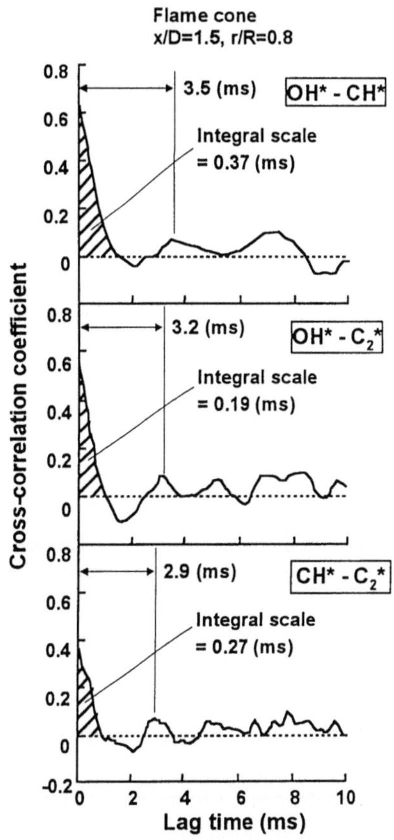

Fig. 11. Cross-correlation coefficients of each chemiluminescence intensity signal such as OH* to CH*, OH* to C_2* and CH* to C_2* at the flame-front (flame cone).

remains whether or not the chemiluminescence can be used to represent the flame-front structure and its features as considering hydrodynamic effects. But here, it was demonstrated that the newly developed measurement technique could be used to identify the relationship between the flame-front thickness and the turbulent scale. This kind of relationship was also found in our previous result in propane/air premixed turbulent flames (Ikeda et al. 2000). This indicated that the local chemiluminescence measurement technique and this analysis were useful to understand flame structure of hydrocarbon premixed flames.

The relationship between the OH*, CH* and C_2* reactions at the flame-front were statistically investigated. Figure 11 shows the cross-correlation of the chemiluminescences of OH* to CH* (OH*-CH*) and OH* to C_2*(OH*-C_2*).

Each cross-correlation shows an initial peak at lag time $\tau = 0$, which indicates that the reactions of OH*, CH* and C$_2$* occurred at almost the same time in the laminar flamelet. The same features were investigated for turbulent flames in shear flows by Katsuki et al. (1990). The second peak in the cross-correlation functions was observed at τ of 3.5, 3.2 and 2.9 ms for the OH*-CH*, OH*-C$_2$* and CH*-C$_2$* cross-correlations respectively. This shows that the CH* and C$_2$* reactions are related to the OH* reaction with similar time scale, while the CH* and C$_2$* reactions were faster than that of the OH*. This cross-correlation coefficient and its time scale should be affected by flame stoicheometry and hydrodynamic effects. It means those can be good parameters for examining flame structure on Borghi diagram. This interesting parameter should be studied in various conditions of u'/S_L.

Concluding remarks

Local chemiluminescence measurements of OH*, CH* and C$_2$* were made at the flame-front of turbulent methane/air premixed flames to flame-front structure. Time-series and statistical analysis of these emission intensities produced the following conclusions:

1. The Cassegrain optics developed were able to detect the local chemiluminescences of OH*, CH* and C$_2$* at the flame-front. The reaction zone was characterized by the intermittent durations of these chemiluminescences and their inter-arrival times.
2. The flame-front thickness and scale was calculated from the time series data of these chemiluminescences. The measured flame-front thickness calculated using the OH* data was an order of magnitude larger than those found using the CH* and C$_2$* results.
3. Simultaneous measurements of the three local chemiluminescences and LDV provided insight into the relationship between the turbulent and chemical scales, and flame-front movement.
4. Local Da based on chemiluminescece of CH* and C$_2$* were proposed. They can be good markers for flame structure.
5. The PDF of OH* chemiluminescence across the flame-front matched the pattern of bi-modal peaks observed in wrinkled laminar flames.
6. The cross-correlations of OH*-CH* and OH*-C$_2$* showed the same time scale, while the CH*-C$_2$* cross-correlation time scale was much smaller.

References

Akamatsu F, Wakabayashi T, Tsushima S, Katsuki M, Mizutani Y, Ikeda Y, Kawahara N, Nakajima T (1999) Development of Light-Collecting Probe with High Spatial

Resolution Applicable to Randomly Fluctuating Combustion Fields. Meas Sci Technol 10-12:1240

Angermaier S, Hardalupas Y, Orain N (2000) Flame Chemiluminescence for Reaction Rate and Local Equivalence Ratio Measurements. 28th Symp (Int) Combust (WIP-Poster)

Ballal DR (1979) The Structure of a Premixed Turbulent Flame. Proc R Soc Lond A 367:353

Bedat B, Cheng RK (1995) Experimental Study of Premixed Flames in Intense Isotropic Turbulence Combust Flame 100:485

Bushmann A, Dinkelacker F, Schtfer T, Schäfer M, Wolfrum J (1996) Measurement of the Instantaneous Detailed Flame Structure in Turbulent Premixed Combustion. Proc Combust Inst 26:437

Chomiak J (1975) Dissipation Fluctuations and the Structure and Propagation of Turbulent Flames in Premixed Gases at High Reynolds Numbers. Proc Combust Inst 16:1665

Damköhler G, Jahrb Z (1940) Der Einfluss der Turbulenz auf die Flammengeschwindigkeit in Gasgemischen. Z Elektrochem 46:601

Deschamps BM, Smallwood GJ, Prieur DRS, Gülder ÖL (1996) Proc Combust Inst 26:287

Docquier N, Belhalfaoui S, Lacas F, Darabiha N, Rolon J (2000) Experimental and Numerical Study of Chemiluminescnece in Methane/Air High Pressure Flames for Active Control Applications. Proc Combust Inst 28 (in press)

Furukawa J, Harada E, Hirano T (1990) Local Reaction Zone Thickness of a Highly Intensity Turbulent Premixed Flame. Proc Combust Inst 23:789

Furukawa J, Hirano T, Williams FA (1998) Burning Velocities of Flamelets in a Turbulent Premixed Flame. Comb Flame 113:487

Gaydon AG (1974) The Spectroscopy of Flames. Chapmann and Hall, London, pp 13

Hanson RK (1988) Combustion Diagnostics: Planar Imaging Techniques. Proc Combust Inst 22:1986

Ikeda Y, Ichi S, Nakai H, Nakajima T (1998) Local Chemiluminescence Measurement for Flame Propagation Analysis. Proc COMODIA-98, pp 411

Ikeda Y, Kojima J, Nakajima T (2000) Measurement of Local Flame-Front Structure in Turbulent Premixed Flame. Proc Combust Inst 28 (In press)

Karlovitz B, Dennision DW, Wells FE (1951) Investigation of Turbulent Flames. J Chem Phys 19:541

Katsuki M, Mizutani Y, Yasuda T, Kurosawa Y, Kobayashi K, Takahashi T (1990) Local Fine Structure and its Influence on Mixing Process in Turbulent Premixed Flames. Comb Flame 82:93

Kojima J, Ikeda Y, Nakajima T (1999) Measuring Local OH* to Analyze Flame Front Movement in a Turbulent Premixed Flame. 35th AIAA/ASME/SAE/ASEE Joint Propulsion Conference and Exhibit, AIAA paper No 99-2784

Kojima J, Ikeda Y, Nakajima T (2000) Spatially Resolved Measurement of OH*, CH* and C_2* Chemiluminescence in the Reaction Zone of Laminar Methane/Air Premixed Flames. Proc Combust Inst 28 (In press)

Kuo KK (1986) Principles of Combustion. John Wiley & Sons, New York, pp 511

Kuznetsov R, Sabelnikov VA (1990) Turbulence and Combustion. Hemisphere Publishing, New York, pp 242

Najm HN, Paul PH, Mueller CJ, Wyckoff PS (1998) On the Adequacy of Certain Experimental of Flame Burning Rate. Comb Flame 113:312

Nguyen QV, Paul PH (1996) The Time Evolution of a Vortex-Flame Interaction Observed via Plannar Imaging of CH and OH. Proc Combust Inst 26:357

Renfro MW, King GB, Laurendeau NM (2000) Scalar Time-Series Measurements in Turbulent CH4/H2/N2 Nonpremixed Flames: CH. Comb Flame 122:139

Roby RJ, Reaney JE, Johnsson EL (1998) Detection of temperature and equivalence ratio in turbulent premixed flames using chemiluminescence. FACT-Vol.22, Proceedings of the 1998 International Joint Power Generation Conference (ASME 1998) 1:593

Samaniego JM, Egolfopoulos FN, Bowman CT (1995) CO2* Chemiluminescence in Premixed Flame. Comb Sci Technol 109:183

Samaniego JM, Mantel T (1999) Fundamental mechanism in Premixed Turbulent Flame Propagation via Flame-Vortex Interaction Part I: Experiment. Comb Flame 118:537

Summerfield M, Riter SH, Kebely V, Mascolo RW (1955) The Structure and Propagation Mechanism of Turbulent Flames in High-Speed Flow. Jet Propulsion 25:377

Tabaczynski RJ, Ferguson CR (1978) A Turbulent Entrainment Model for Spark-Ignition Engine Combustion. SAE Paper No 770647

Tsushima S, Saitoh H, Akamatsu F, Katsuki M (1998) Observation of Combustion Characteristics of Droplet Clusters in a Premixed-Spray Flame by Simultaneous Monitoring of Planar Spray Images and Local Chemiluminescece. Proc Combust Inst 27:1967

Walsh KT, Long MB, Tanoff MA, Smooke MD (1998) Experimental and Computational Study of CH, CH*, and OH* in an Axisymmetric Laminar Diffusion Flame. Proc Combust Inst 27:615

Williams FA (1985) Combustion Theory, 2nd ed. Addison-Wisley, Redwood City, pp 156

Yoshida A, Kakinuma H, Kotani Y (1992) Structure of Highly Turbulent Premixed Flames. Proc Combust Inst 24:397

Yoshida A, Tsuji H (1982) Characteristic Scale of Wrinkles in Turbulent Premixed Flames. Proc Combust Inst 19:403

Sound Generation in Chemically Reacting Mixing Layers

Toshio Miyauchi, Mamoru Tanahashi and Ye Li

Department of Mechanical and Aerospace Engineering, Tokyo Institute of Technology,
2-12-1 Ookayama, Meguro-ku, Tokyo 152-8552, Japan

Summary. Direct numerical simulations have been performed to clarify the sound generation mechanism in a two-dimensional chemically reacting compressible mixing layers. The effects of heat release on the mechanism of sound generation are investigated. The pressure fluctuations generated in the reacting mixing layers with heat release are significantly larger than that in the case without heat release, which suggests that the characteristics of sound are mainly determined by the heat release. The acoustic source term in Lighthill's equation is dominated by entropy component in the case with heat release, while it is governed by Reynolds stress component in the case without heat release.

The far-field sound computed by DNS is compared with the predictions based on the acoustic analogies. In the case with heat release, the acoustic analogies proposed by Lighthill and Powell fail to predict the far-field sound when the source size is selected to be $2\varLambda$. However, the pressure fluctuation in the far field predicted by Lighthill's analogy shows good agreement with the DNS result when the source size is selected to be $3\varLambda$. A new analogy including Powell's acoustic source term and entropy term is proposed, which can predict the far-field sound excellently for both cases of the source size $2\varLambda$ and $3\varLambda$.

Key words. DNS, Noise, Reacting Flow, Heat Release, Acoustic Analogy

Introduction

In the design process of high efficiency combustor, it is important to reduce the combustion noise and to inhibit the combustion-driven oscillations. Combustion noise is produced by the non-steady oscillation of flame, which is due to the interaction between the vortex structures and flame in the shear flow. Combustion oscillation is enhanced by the resonance of combustion noise in combustor. In order to understand and control these phenomena, it is necessary to make clear the mechanism of sound generation in chemically reacting flows.

The researches on the sound generation can be divided into three groups. The first approach is using acoustic analogy. It includes the Reynolds stress acoustic model proposed by Lighthill (1952), the vortex acoustic model by Powell (1964) and the acoustic model proposed by Möhring (1978). The second approach is experimental one. Laufer et al. (1983) have conducted the experiment on sound generation from low Mach number jet, and investigated the relation between the sound source and the pairing process of large-scale structures in shear layer. With the recent development of high-speed and large storage computer, analysis of the sound generation by direct numerical simulations (DNS) becomes possible. Ho et al. (1988) have investigated the sound generation in two-dimensional temporally evolving mixing layer. Colonius et al. (1994), Mitchell et al. (1995) have performed the DNS of sound generation by compressible vortex. The far field sound was also predicted using acoustic analogies and compared with the DNS data. Colonius et al. (1997) have also clarified the sound generation in two-dimensional spatially evolving mixing layer.

In this study, we conduct the DNS of two-dimensional chemically reacting compressible mixing layers to clarify the sound generation mechanism. We focus on the determination of principal acoustic source term, the relation between sound generation and chemical reaction and the effects of heat release on the sound generation. Moreover, the far-field sound can be predicted by the acoustic analogy. The results predicted by acoustic analogies are compared with the results of DNS.

DNS of chemically reacting mixing layer

The external forces, Soret effect, Dufour effect, pressure gradient diffusion, bulk viscosity and radiative heat transfer are assumed to be negligible in this work. The chemical reaction is idealized to be a single step, irreversible reaction with heat release ($A+B \rightarrow P+\Delta H$). The flow field is governed by the mass, momentum, energy and species conservation equations, and the equation of state. The governing equations are non-dimensionalized by the density and temperature of the free-stream flow, the difference of free-stream velocities, and the initial vorticity thickness.

In this study, temporally developing mixing layers are analyzed by DNS. Periodic boundary condition is used in streamwise (x) direction and non-reflecting boundary condition (NSCBC, Poinsot et al. 1992) is applied in transverse (y) direction. The governing equations are discretized by spectral method in the x direction and by the fourth order central finite difference scheme in the y direction. Aliasing errors from nonlinear terms are fully removed by 3/2 rule in the x direction. A second-order Adams-Bashforth method is used to advance the equations in time. The computational domain in the x and y direction is selected to be 2Λ and 8Λ respectively, where Λ is the most unstable wavelength for the initial mean velocity profile. The simulation is performed on the mesh of

128×1025 for the case of Re=400, Mc=0.2, Pr=0.7, Sc=0.7 and γ=1.4. In order to investigate the

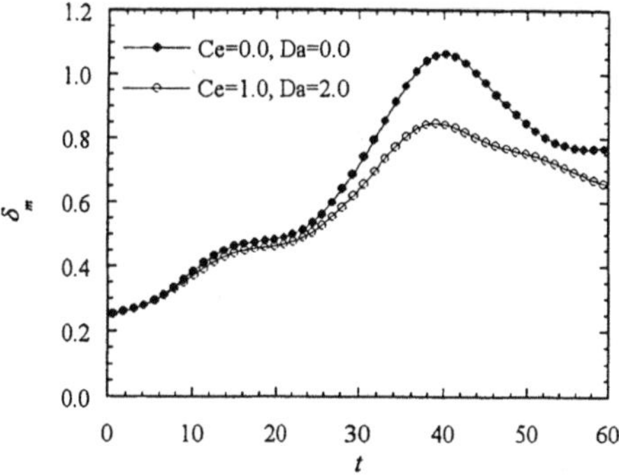

Fig. 1. Development of momentum thickness.

effects of heat release, two cases of Ce=1.0, Da=2.0 (with heat release) and Ce=0.0, Da=0.0 (without heat release) are computed. The initial velocity field is composed of a hyperbolic tangent velocity profile and perturbations which contain the most unstable fundamental mode and its subharmonic.

Mechanism of sound generation in chemically reacting mixing layer

In this section, the DNS results are presented to clarify the mechanism of sound generation in chemically reacting mixing layers. Figure 1 shows the development of momentum thickness defined by

$$\delta_m = \frac{1}{\rho_0 \Delta U^2} \int_{-Ly/2}^{Ly/2} \langle \rho \rangle \left(U_1 - \langle u \rangle \right) \left(\langle u \rangle - U_2 \right) dy , \tag{1}$$

where ρ_0 represents the density of free-stream flow, U_1, U_2 and ΔU represent the velocities of high-speed side, low-speed side and the difference of these two respectively. In the mixing layer without heat release, the development of momentum thickness corresponds to the development of large scale structures. Momentum thickness grows up rapidly in the process of vortex roll-up (t<15), then reaches the plateau, and increases again in the period of pairing (25<t<40). In the case with heat release, the growth rate of momentum thickness is inhibited. The time corresponding to the peak of momentum thickness becomes a little bit earlier due to the effects of heat release.

Figure 2 shows the temporal evolution of the pressure fluctuations in the far field. The measurement point is located at 3Λ from the center of shear layer. The time coordinate is shifted to adjust the sound propagating from the center to the measurement point. The pressure fluctuation with relatively low amplitude and

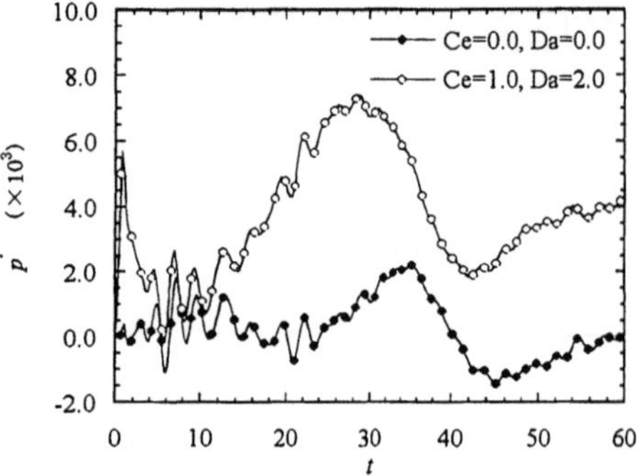

Fig. 2. Temporal evolution of the pressure fluctuations in the far field.

high frequency is generated in the process of vortex roll-up, while the amplitude of pressure fluctuation increases significantly in the period of pairing. Moreover, the maximum amplitude of pressure fluctuation generated in the mixing layer with heat release is about four times larger than that without heat release. Similar to the variation of momentum thickness, time corresponding to the peak of pressure fluctuation becomes earlier in the case with heat release.

Figure 3 show the contour plots of dilatation ($d=\nabla\cdot\boldsymbol{u}$) at $t=10$ and $t=40$. The solid lines represent the positive values and the dash lines represent the negative values in these figures. For the case without heat release, the sound source shows quadrupole character in the process of vortex roll-up ($t=10$), while in the pairing process ($t=40$) of two large-scale structures, two quadrupole merger to one large quadrupole. For the case with heat release, the heat release rate is relatively low in the early stage of vortex roll-up, and shows weak effects on the sound generation. As a result, the sound source shows quadrupole character similar to that without heat release. However, the sound source can not be approximated by quadrupole in the period of pairing, because the heat release rate increases significantly.

As an acoustic analogy, Lighthill (1952) has rearranged the exact continuity and momentum equations into a wave equation with a source term on the right-hand side as follow:

$$\frac{\partial^2 \rho'}{\partial t^2} - \frac{1}{M^2}\frac{\partial^2 \rho'}{\partial x_i \partial x_i} = \frac{\partial^2 T_{ij}}{\partial x_i \partial x_j} , \tag{2}$$

where T_{ij} is the Lighthill's turbulent stress tensor defined by

$$T_{ij} = \rho u_i u_j + \frac{1}{M^2}\delta_{ij}\left(\frac{1}{\gamma}p - \rho\right) - \frac{1}{Re}\tau_{ij} , \tag{3}$$

in which the first, second and third term on the right hand side represent the Reynolds stress (T_R), entropy (T_E) and viscous (T_V) component respectively.   In order

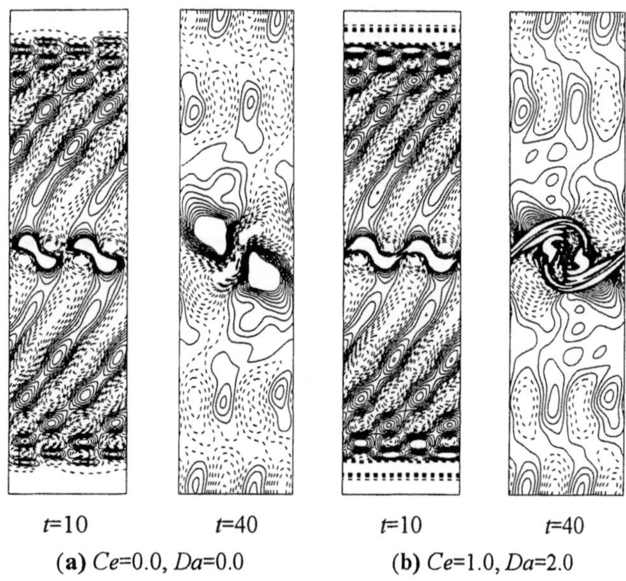

Fig. 3. Contour plots of dilatation.

to evaluate the magnitude of each acoustic source term, the integration of acoustic source terms defined by

$$<< T_k >> = \int_{-Ly/2}^{Ly/2} \int_0^{Lx} T_k^2 \, dxdy , \qquad (4)$$

is used in this study.

Figure 4 show the contributions of each acoustic source terms.   For the case without heat release, the Reynolds stress component is dominant and shows two peaks at the times corresponding to the vortex roll-up and pairing.   However, in the case with heat release, the entropy component is significantly larger than the other two components.   The entropy component shows the peak at the time between the vortex roll-up and paring.   The viscous component is negligible for the two cases.   It can be concluded that the sound generation is governed by the variation of vortex in the case without heat release, while the sound generation is due to the chemical reaction in the case with heat release.

Figure 5 shows the development of total heat release rate define as follow:

$$Q = CeDa \iint \rho^2 Y_A Y_B \, dxdy . \qquad (5)$$

The total heat release rate shows the peak in the period between vortex roll-up and pairing.   Temporal evolution of the total heat release rate shows similar profile with the variation of total acoustic source term (Fig. 4(b)).   The variation of total heat release rate is also consistent with the pressure fluctuation in the far-field (Fig. 2). Accordingly, it is concluded that the sound generation in the reacting mixing layer is determined by the heat release rate.

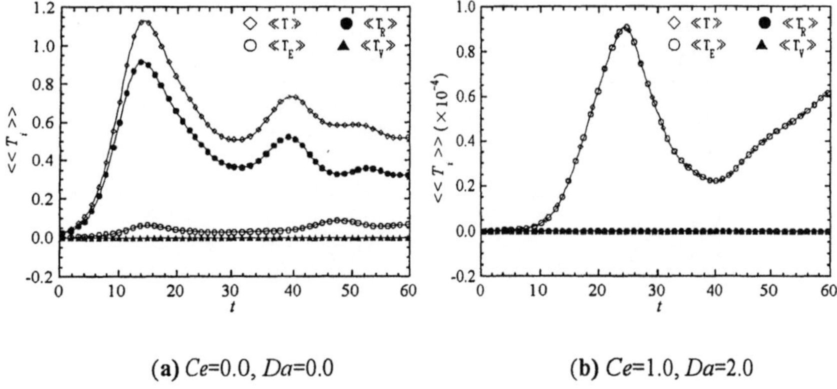

(a) Ce=0.0, Da=0.0   (b) Ce=1.0, Da=2.0

Fig. 4. Contributions of acoustic source terms.

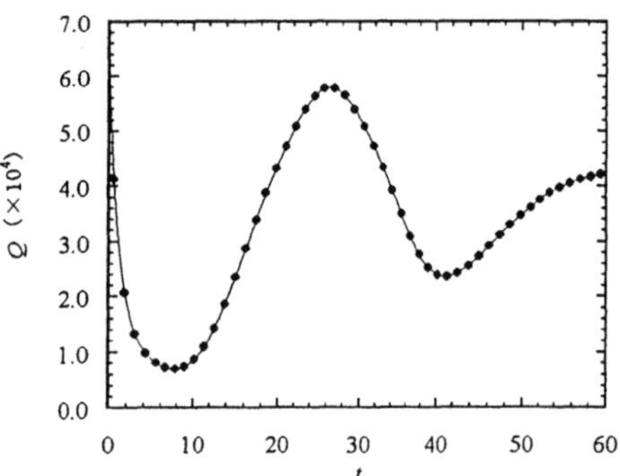

Fig. 5. Temporal evolution of total heat release, Ce=1.0, Da=2.0.

Here, the relation between entropy component and total heat release rate will be discussed. In the case of perfect gas, if the variations of all quantities are assumed to be small, the continuity and energy conservation equations can be linearised to the following two equations (Lighthill 1978, Tanahashi et al. 1995).

$$\frac{\partial \rho'}{\partial t} = -\rho_0 d , \tag{6}$$

$$\frac{\partial p'}{\partial t} = -\gamma p_0 d + q , \tag{7}$$

where q represents the total heat release rate. From the above linearised energy equation (Eq. (7)), it can be seen that the variation of pressure fluctuation can be related to the total heat release rate. By using these two equations and the definition of entropy component, the entropy component can be expressed by the following equation.

$$T_E = \frac{\partial^2}{\partial x_i x_j}\left[(p - p_0) - c^2(\rho - \rho_0)\right] = \frac{\partial^2}{\partial x_i x_j}\left(\frac{q}{\rho_0 d}\right). \tag{8}$$

The above equation shows that the entropy component becomes very small in the case without heat release. However, if the chemical reaction release heat, the entropy component is determined by the second spatial derivative of total heat release rate.

Prediction of far-field sound using acoustic analogies

In this section, the pressure fluctuations in the far-field are predicted by using acoustic analogies and compared with the result of DNS. The Lighthill's equation clearly shows that the sound is generated in the turbulent flow. However, it is unclear how the sound is generated from the flow field. In order to clarify this problem, Powell (1964) has proposed a theory of vortex sound and deduced the following wave equation.

$$\frac{\partial^2 \rho'}{\partial t^2} - \frac{1}{M^2}\nabla^2 \rho' = \nabla\cdot(\rho\,\omega\times u). \tag{9}$$

This equation shows that the variation of vorticity plays an important role in the sound generation. In fact, the relation between Powell's acoustic source terms and Ligthill's acoustic source terms can be expressed as:

$$T = \nabla\cdot\left[\rho\,\omega\times u + \nabla\left(\tfrac{1}{2}\rho u^2\right) - \tfrac{1}{2}u^2\nabla\rho + u\nabla\cdot(\rho u)\right] + T_R + T_V. \tag{10}$$

It can be seen that the Powell's acoustic source term is a part of the Lighthill's source term.

The wave equations are discretized by the spectral method in the x direction and fourth-order central finite difference scheme in the y direction. A second-order Adams-Bashforth scheme is used to advance the wave equations in time. Periodic boundary condition is applied in the x direction and non-reflecting boundary conditions proposed by Engquist et al. (1979) are used in the y direction. The computational grids are same as those used in the DNS.

When the wave equations are used to predict the far-field sound, the size of acoustic source used in the prediction is a important parameter. In the near-field, the wave equations with the acoustic source term obtained by the DNS are solved. However, in the far-field, the wave equations which express the propagation of sound wave are solved without acoustic source. The source size is defined as the length of near-field in the y direction.

Figure 6 shows the pressure fluctuations in the far filed predicted by Lighthill's acoustic analogy (Eq. (2)) and Powell's acoustic analogy (Eq. (9)) for the case without heat release. The DNS result is also plotted for comparison. In the case where the source size is selected to be 2Λ, the pressure fluctuations predicted by Lighthill's analogy and Powell's analogy is nearly consistent with the result of DNS in the process of vortex roll-up. In the period of pairing, the pressure

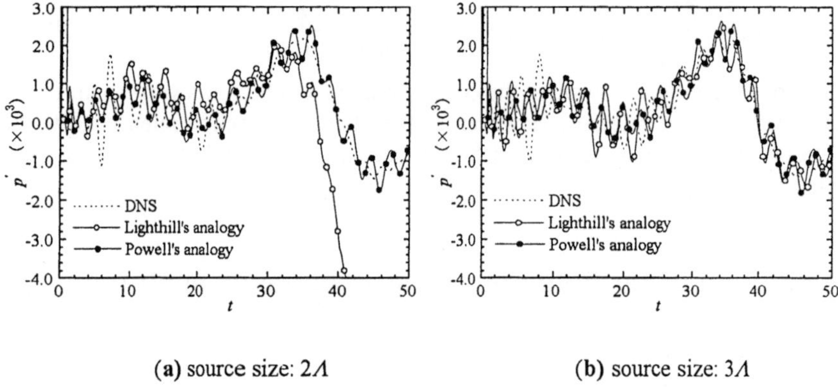

(a) source size: 2Λ (b) source size: 3Λ

Fig. 6. Comparison of pressure fluctuation obtained by DNS and predicted by acoustic analogies, $Ce=0.0, Da=0.0$.

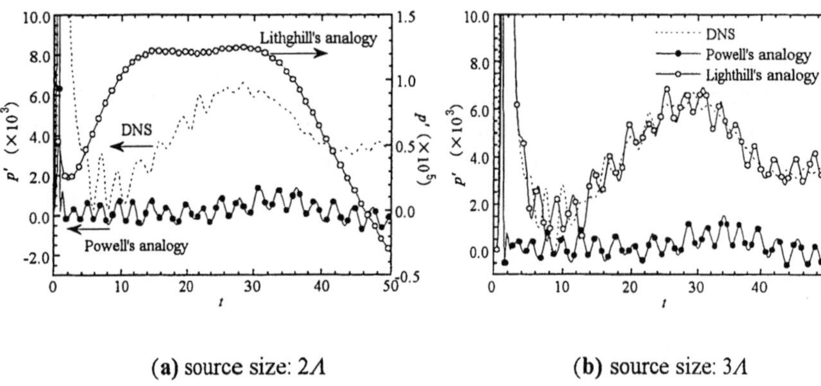

(a) source size: 2Λ (b) source size: 3Λ

Fig. 7. Comparison of pressure fluctuation obtained by DNS and predicted by acoustic analogies, $Ce=1.0, Da=2.0$.

fluctuation predicted by Powell's analogy shows good agreement with the DNS result,while the prediction by Lighthill's analogy shows very low level of pressure fluctuation compared with the DNS. However, if the source size is selected to be 3Λ, the pressure fluctuations predicted by Lighthill's acoustic analogy and Powell's acoustic analogy is almost same, which show good agreement with the DNS result. It suggests that 2Λ of source size is enough to represent the Powell's acoustic source, while 3Λ is necessary to represent the Lighthill's acoustic source.

Figure 7 shows the pressure fluctuations in the far filed obtained by DNS and predicted by Lighthill's analogy and Powell's analogy for the case with heat release. The pressure fluctuation predicted by Powell's acoustic analogy is smaller than that of DNS for both cases where the source size is selected to be 2Λ and 3Λ.

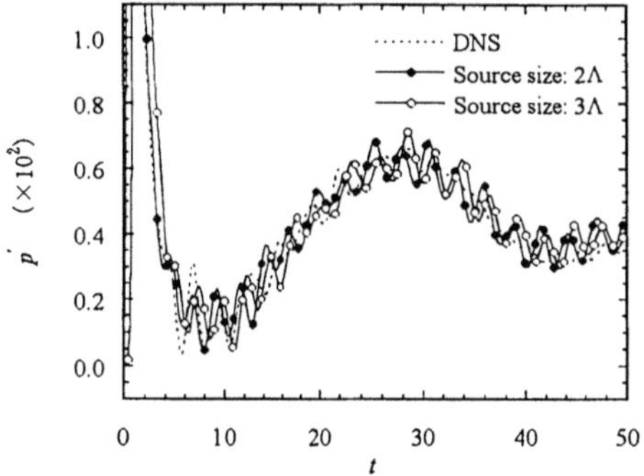

Fig. 8. Comparison of pressure fluctuation obtained by DNS and predicted by proposed acoustic analogy, Ce=1.0, Da=2.0.

If the source size is selected to be 2Λ, the pressure fluctuation predicted by Lighthill's acoustic analogy is much larger than that of DNS. However, if the source size is selected to be 3Λ, the pressure fluctuation predicted by Lighthill's acoustic analogy shows good agreement with the DNS result. This suggests that the source size shows very large effect on the pressure fluctuations in the far field when Lighthill's analogy is used to predict the reacting mixing layers.

As shown in Fig. 4(b), the entropy component is much larger than the Reynolds stress component in the case with heat release. Therefore, it is expected that the entropy component plays an important role in the sound generation. In this work, we propose the following acoustic analogy equation with the acoustic source terms including Powell's acoustic term and entropy component.

$$\frac{\partial^2 \rho'}{\partial t^2} - \frac{1}{M^2}\nabla^2 \rho' = \nabla \cdot \left(\rho\,\omega\times u\right) + T_E \tag{11}$$

Figure 8 shows the pressure fluctuations in the far filed predicted by the proposed acoustic analogy. The pressure fluctuation predicted by proposed acoustic analogy shows good agreement with the DNS for the two cases where the source size is selected to be $2\,\Lambda$ and $3\,\Lambda$ respectively. This suggests that the proposed analogy can predict the far-field sound excellently even the small source size is used.

Conclusions

In this study, DNS have been carried out to clarify the mechanism of sound generation in the chemically reacting mixing layers and the following conclusions are obtained.

(1) The pressure fluctuations generated in the reacting mixing layer with heat release are significantly larger than that without heat release. The acoustic source term in Lighthill's equation is dominated by entropy component and the Reynolds stress component is considerably small in the case with heat release, while it is governed by the Reynolds stress component in the case without heat release. The viscous component is negligible in both cases.

(2) For the case without heat release, the far-field sound predicted by Lighthill's and Powell's acoustic analogy show good agreement with the DNS result when the source size is selected to be 3Λ. If the source size is selected to be 2Λ, Powell's analogy can predict far-field sound excellently, while the pressure fluctuation predicted by Lighthill's analogy is small compared with the DNS.

(3) For the case with heat release, the far-field sound predicted by Powell's analogy is very small compared with the DNS. The far-field sound by Lighthill's analogy is significantly large compared with the DNS result when the source size is selected to be 2Λ. However, the pressure fluctuation in the far field predicted by Lighthill's analogy shows good agreement with the DNS when the source size is selected to be 3Λ.

(4) A new analogy including Powell's acoustic source term and entropy component is proposed, which can predict the far-field sound excellently for both cases where the source size is selected to be 2Λ and 3Λ.

References

Colonius T, Lele S K, Moin P (1994), The scattering of sound waves by a vortex: numerical simulations and analytical solutions. J Fluid Mech 260: 271-298

Colonius T, Lele S K, Moin P (1997), Sound generation in a mixing layer. J Fluid Mech 330: 375-409

Engquist B and Majda A (1979), Radiation boundary conditions for acoustic and elastic wave calculations, Commun Pure Applied Math 23: 313-357

Ho C M, Zohar Y, Moser R D, Roger M M, Lele S K, Buell J C (1988), Phase decorrelation, streamwise vortices and acoustic radiation in mixing layers. Center for Turbulence Research, Proceedings of the Summer Program 1988: 29-39

Laufer J, Yen T C (1983), Noise generation by a low-Mach-number jet. J Fluid Mech 134: 1-31

Lighthill J (1978) Waves in Fluids. Cambridge University Press, pp 1-5

Lighthill M J (1952), On sound generated aerodynamically I. General theory. Proc R Soc Lond A 211: 564-587

Möhring W (1978), On vortex sound at low Mach number. J Fluid Mech 85 part4: 685-691

Mitchell B E, Lele S K, Moin P (1995), Direct computation of the sound from a compressible co-rotating vortex pair. J Fluid Mech 285: 181-202

Poinsot T J, Lele S K (1992), Boundary conditions for direct simulations of compressible viscous flows. J Comput Phys 101: 104-129

Powell A (1964), Theory of vortex sound. J Acoust Soc Am 36: 177-195

Tanahashi M, Miyauchi T (1995), Interaction between turbulence and chemical reaction in turbulent diffusion flames. Proceeding of the ASME/JSME, Thermal Engineering: 105-110

Turbulent Transport Properties of Wrinkled Flames

Shinnosuke Nishiki[1], Tatsuya Hasegawa[1], and Ryutaro Himeno[2]

[1]Department of Environmental Technology and Urban Planning, Graduate School of Engineering, Nagoya Institute of Technology, Gokiso-cho, Showa-ku, Nagoya 466-8555, Japan
[2]Computer and Information Division, The Institute of Physical and Chemical Research, 2-1 Hirosawa, Wako-shi, Saitama 351-0198, Japan

Summary. Direct numerical simulation of a turbulent premixed flame with a single-step irreversible Arrhenius-type reaction is performed at $u'/u_L = 0.88$, $l_t/\delta = 15.9$ and $\rho_u/\rho_b = 7.53$, and turbulent transport properties are evaluated. A fully developed stationary wrinkled flame is obtained in a domain of 8 mm × 4 mm × 4 mm. Turbulent fluctuations generally increase in the flame region, but streamwise component increases more than transversal components. This results in the generation of anisotropic turbulence in the flame region. Analysis based on the Favre-averaged transport equation of turbulent kinetic energy shows that pressure gradient term and pressure work term increase turbulent kinetic energy in the flame region, while diffusion and dissipation term and velocity gradient term decrease it in the flame region. The counter-gradient diffusion dominates turbulent scalar flux in the DNS data, and the BML model well predicts it. Analysis based on the Favre-averaged transport equation of turbulent scalar flux shows that mean pressure gradient term, velocity-reaction rate correlation term and fluctuating pressure term play important roles on production of counter-gradient diffusion, while mean velocity gradient term, mean progress variable gradient term and dissipation terms suppress it.

Key words. Wrinkled Flames, DNS, Flame-Generated Turbulence, Counter-Gradient Diffusion

Introduction

Numerical simulation becomes to play an important role in design of practical combustors, but turbulent combustion models are still required even by using modern supercomputers. The structure and characteristics of turbulent premixed flames are progressively made clear experimentally, for example, by Kobayashi et al. (1997, 1998), Furukawa et al. (2000). However, it is impossible to measure all variables in time and in space experimentally. Therefore DNS takes an important place in order to evaluate and develop turbulent combustion models.

There are a few databases of three-dimensional turbulent premixed flames with a simple reaction. For example, Trouvé and Poinsot (1994) calculated a propagating premixed flame in a decaying turbulence with one-step irreversible reaction, and studied the effect of Lewis numbers on statistical properties such as the turbulent burning velocity and the local flame structure. The initial characteristics of turbulence are $u'/u_L = 10.0$ and $l_t/\delta = 5.2$ and the density ratio of the flame is 4.0. Rutland and Cant (1994) also calculated a stationary premixed flame in an incoming turbulence using one-step irreversible reaction and low Mach number approximation. The initial characteristics of turbulence are $u'/u_L = 1$ and $l_t/\delta = 30$ and the density ratio of the flame is 3.3. Besides these simple reaction cases, Tanahashi et al. (1999) performed three-dimensional DNS of H_2-air turbulent premixed flames with detailed kinetic mechanism including 12 reactive species and 27 elementary reactions. The initial characteristics of turbulence are $u'/u_L = 3.0$ and $l_t/\delta = 1.74$.

Hasegawa et al. (1999) and Nishiki et al. (2000) studied turbulence, flame structure and transport properties of turbulent premixed flames on the basis of DNS with one-step irreversible reaction, though flames were not enough developed. The initial characteristics of turbulence are $u'/u_L = 4.8$ and $l_t/\delta = 8.0$ and the density ratio of the flame is 7.53. In this study, a fully developed stationary wrinkled flame is obtained in a turbulent flow of $u'/u_L = 0.88$ and $l_t/\delta = 15.9$ with a flame of $\rho_u/\rho_b = 7.53$. In the following sections, simulation method will be described and evolutions of turbulence and turbulent scalar flux through the flame are discussed.

Direct numerical simulation

Basic equations

Following assumptions are used in the simulation: 1) The chemical reaction is a single-step irreversible one with heat release, where the molecular weights of reactants and products are the same. 2) The bulk viscosity, the Soret and the Dufour effects, and the pressure gradient diffusion are neglected. 3) The specific heat at constant pressure and the specific heat ratio are constant. 4) The equation of state of the burned and unburned gases is that of an ideal gas. Following the above assumptions, basic equations are written as follows:

Mass conservation

$$\frac{\partial \rho}{\partial t} + \frac{\partial (\rho u_j)}{\partial x_i} = 0$$

Momentum conservation

$$\frac{\partial(\rho u_i)}{\partial t}+\frac{\partial(\rho u_i u_j)}{\partial x_j}+\frac{\partial p}{\partial x_i}=\frac{\partial \tau_{ij}}{\partial x_j}\qquad (i=1,2,3)$$

Energy conservation

$$\frac{\partial e_t}{\partial t}+\frac{\partial\{(e_t+p)u_j\}}{\partial x_j}=\frac{\partial(u_j \tau_{kj})}{\partial x_k}-\frac{\partial q_j}{\partial x_j}$$

Species conservation

$$\frac{\partial(\rho Y)}{\partial t}+\frac{\partial(\rho Y u_j)}{\partial x_j}=\frac{\partial}{\partial x_j}\left(\rho D\frac{\partial Y}{\partial x_j}\right)+W$$

Stress tensor

$$\tau_{ij}=\mu\left(\frac{\partial u_i}{\partial x_j}+\frac{\partial u_j}{\partial x_i}-\frac{2}{3}\delta_{ij}\frac{\partial u_k}{\partial x_k}\right)$$

Total energy

$$e_t=\rho\,QY+\frac{\rho RT}{\gamma-1}+\frac{\rho}{2}\left(u^2+v^2+w^2\right)$$

Heat flux

$$q_i=-\lambda\,\frac{\partial T}{\partial x_i}-\rho DQ\frac{\partial Y}{\partial x_i}$$

Reaction term

$$W=-B\rho YT^{\beta}\exp\left(-\frac{\theta}{T}\right)$$

where μ,λ,D are the viscosity, the thermal conductivity and the diffusion coefficient, Q is the released heat, B is the frequency factor, β is the index of temperature dependency and θ is the characteristic temperature of the activation energy. Viscosity coefficient is evaluated by the 0.7 power of temperature, while the Lewis number is 1.0 and the Prandtl number is 0.75.

Properties of premixed gas

A premixed gas is supposed to have the pressure of 0.1 MPa, the initial temperature of 300 K, the density of 0.942 kg/m^3, the released heat Q of 2.45×10^6 J/kg, the specific heat at constant pressure C_p of 1.25×10^3 J/kg·K, the specific heat ratio γ of 1.4, and the heat conductivity λ of 7.42×10^{-2} W/m·K. Thus the adiabatic temperature T_a is 2260 K and the density ratio of the flame ρ_u/ρ_b is 7.53. The index of temperature dependency β is 1.0, the characteristic temperature of activation energy θ is 19600 K and the laminar burning velocity u_L is set 0.600 m/s. As a result, the frequency factor B is 2.79×10^6 1/s·K and the flame thickness δ is 0.217 mm.

Calculation method

The simulation domain is a box of 8 mm × 4 mm × 4 mm. The x coordinate is taken in the streamwise direction and y, z coordinates are taken in the transversal directions as shown in Fig. 1.

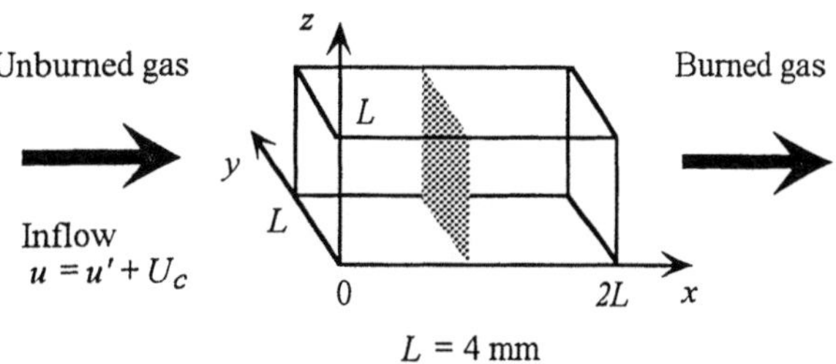

Fig. 1. Simulation domain and coordinate system.

Boundary conditions in y, z directions are periodic, and those in x direction are non-periodic to be able to treat incoming flesh turbulent mixture and outgoing burned gas. Inadequate boundary conditions for incoming and outgoing flows may produce reflecting pressure waves and false vorticity. Thus the NSCBC (Navier-Stokes Characteristic Boundary Conditions) proposed by Poinsot and Lele (1992) and Baum et al. (1994) is applied to the boundaries of the x direction. At the inflowing boundary, the velocity is assigned by assuming Taylor's hypothesis and shifting the phase of a homogeneous isotropic turbulence calculated beforehand. A homogeneous isotropic turbulence is calculated by giving the following energy spectrum to satisfy incompressibility:

$$E(k) = 16u'^2_0 \left(\frac{2}{\pi}\right)^{\frac{1}{2}} \frac{k^4}{k_0^5} \exp\left\{-2\left(\frac{k}{k_0}\right)^2\right\}$$

where $k_0 = 6$ and $u'_0 = 3.87$ m/s. The characteristics of the obtained homogeneous isotropic turbulence after about 2 times the turnover time are listed in Table 1. At the inflowing boundary, this homogeneous isotropic turbulence comes into the domain with a mean velocity of 0.60 m/s by the phase shifting.

Table 1. Characteristics of homogeneous isotropic turbulence

u'/u_L	l_t/δ	l_m/δ	l_d/δ	Re_t
0.88	15.9	9.44	0.65	95.5

u': Turbulent intensity, u_L: Laminar burning velocity = 0.60 m/s,
l_t : Integral length scale, l_m : Taylor micro scale, l_d : Kolmogorov scale,
δ : Flame thickness = 0.217 mm, Re_t: Reynolds number based on the integral length scale.

The 6th-order central finite difference method is used for the x direction to treat non-periodic boundary conditions and the Fourier spectral collocation method is used for y, z direction. The number of grid points is 512 in the x direction and 128 in y, z directions. The 3rd-order Runge-Kutta method is used for time integration.

A vector-parallel computer (Fujitsu VPP 700) with 32 PEs is used for this direct numerical simulation. CPU time is about 50 hours for 1ms of real-time simulation.

Initial conditions

The homogeneous isotropic turbulence listed in Table 1 is assigned in the domain with a mean inflowing velocity of 0.60 m/s. Distributions of temperature, pressure, mass fraction and expansion velocity are also given as if a planar flame is located in the simulation domain as shown in Fig. 1. The mean inflowing velocity is changed on the way of the calculation to keep a developed wrinkled flame in the simulation domain.

Results and discussions

Wrinkled flame

The mean inflowing velocity U_c was changed from 0.60 m/s to 1.0 m/s at $t = 4.65$ ms and from 1.0 m/s to 1.2 m/s at $t = 9.30$ ms on the way of the calculation. The contour surface of temperature drawn at 1470 K at $t = 13.44$ ms is shown in Fig. 2. It is found that a stationary wrinkled flame appears in the turbulent flow. The flame region exists between $x/L = 0.25$ and 1.07, which is about 15 times larger than the laminar flame thickness. The contour surface of total vorticity drawn at 0.29×10^4 1/s is also shown in Fig. 2. Fine scale structure of turbulence exists in the incoming turbulent flow and vorticity production occurs near the wrinkled flame, though the mechanism has not been cleared yet.

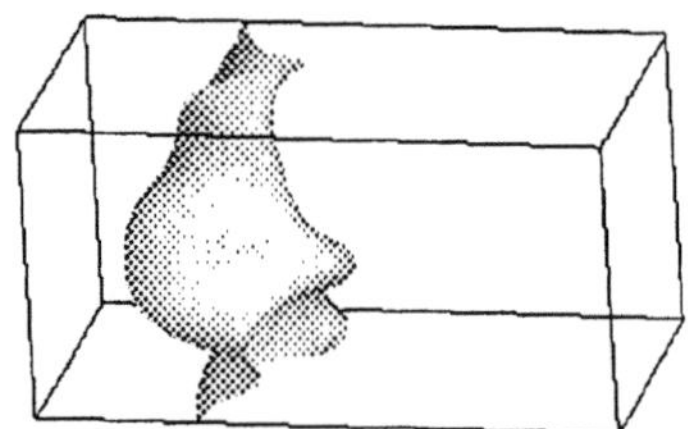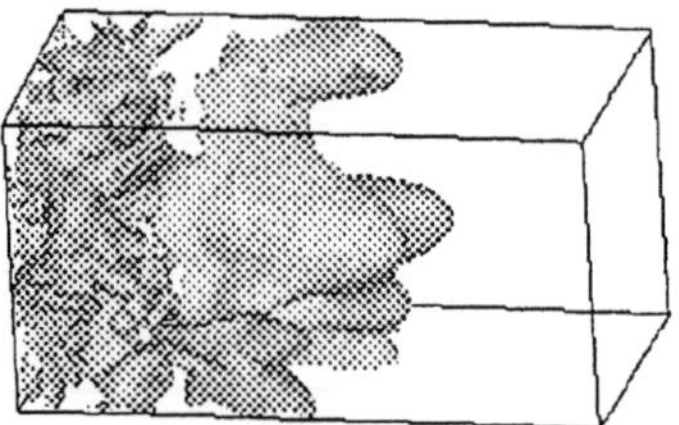

Fig. 2. Contour surfaces at $t = 13.44$ ms.
Left side: Temperature at 1470K [$0.75(T_a\text{-}T_0)$].
Right side: Total vorticity at 0.29×10^4 1/s [$0.15|\omega|_{max}$].

Turbulent burning velocity

Temporal evolution of turbulent burning velocity is shown in Fig. 3. The turbulent burning velocity is defined as follows (Trouvé and Poinsot 1994):

$$u_T = -\frac{1}{\rho_u Y_u A_L}\int_V W dV \ ,$$

where ρ_u is the density of the unburned gas, Y_u is the fuel mass fraction in the unburned gas ($Y_u = 1$), A_L is the area of the initial flame and W is the reaction rate. Mean turbulent burning velocity is 1.146 m/s after $t = 9.30$ ms, which is very close to $u_L + u' = 1.128$ m/s.

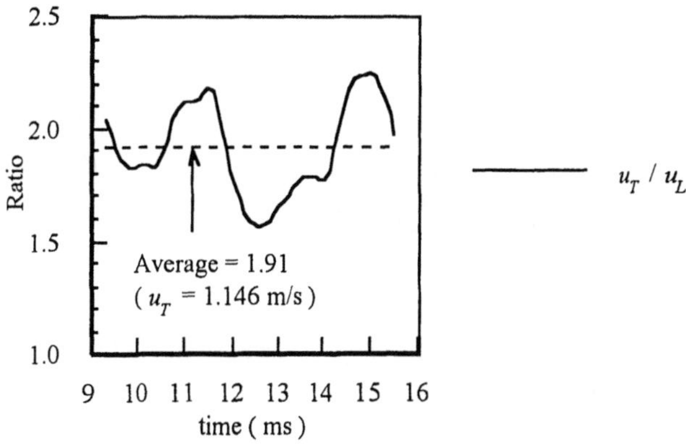

Fig. 3. Temporal evolution of turbulent burning velocity.

Generation of turbulent kinetic energy

Streamwise evolution of turbulent fluctuations at $t = 13.44$ ms is shown in Fig. 4. All fluctuations slightly decay in front of the flame region, but they increase in the flame region. Especially streamwise component increases more than twice as much as transversal components in the flame region and this results in the generation of anisotropic turbulence in the flame region.

The mechanism of turbulence generation is analyzed on the basis of the Favre-averaged transport equation of turbulent kinetic energy along streamwise direction. The transport equation is written as follows:

$$\frac{\partial \tilde{k}}{\partial t} + \tilde{u}_k \frac{\partial \tilde{k}}{\partial x_k} = -\underbrace{\frac{\overline{\rho u_i'' u_k''}}{\bar{\rho}}\frac{\partial \tilde{u}_i}{\partial x_k}}_{(I)} -\underbrace{\frac{1}{\bar{\rho}}\frac{\partial \overline{\rho u_i'' u_i'' u_k''}}{2\partial x_k}}_{(II)} -\underbrace{\frac{\overline{u_i''}}{\bar{\rho}}\frac{\partial \bar{p}}{\partial x_i}}_{(III)} -\underbrace{\frac{1}{\bar{\rho}}\overline{u_i''\frac{\partial p'}{\partial x_i}}}_{(IV)} +\underbrace{\frac{1}{Re}\frac{1}{\bar{\rho}}\overline{u_i''\frac{\partial \tau_{ik}}{\partial x_k}}}_{(V)}$$

where term (I) is the production due to velocity gradient, term (II) is the turbulent diffusion, term (III) is the production due to pressure gradient, term (IV) is the pressure work and term (V) is the diffusion and the dissipation. Evolution of each term, which is averaged between $t = 9.30$ ms and $t = 15.50$ ms, is shown in Fig. 5. As shown in Fig. 5, pressure gradient term and pressure work term produce kinetic energy in the flame region, while diffusion and dissipation term and velocity gradient term decrease it.

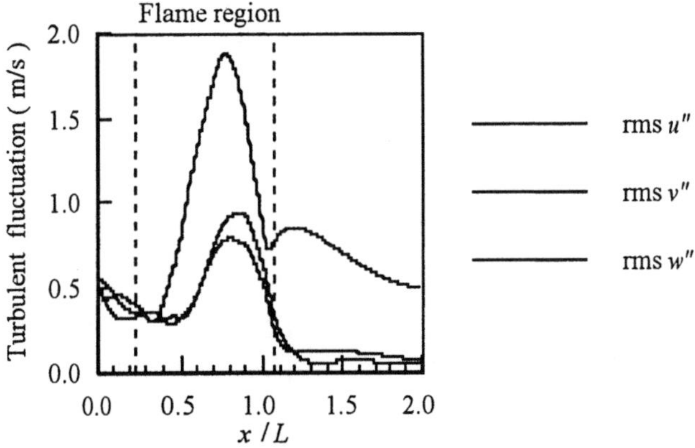

Fig. 4. Streamwise evolution of turbulent fluctuation.

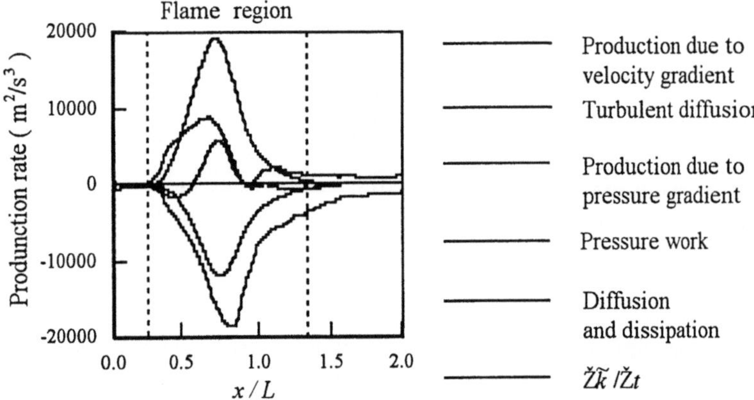

Fig. 5. Streamwise balance of the production rate of turbulent kinetic energy. Flame region is defined between 0.01 and 0.99 of averaged progress variable. $\partial \tilde{k}/\partial t$ is calculated from the transport equation by subtracting the convection term from the right hand side.

The pressure work term can be decomposed into the pressure diffusion term and the pressure dilatation term. It is shown that the pressure dilatation term always takes positive value, but the pressure diffusion term changes its sign from positive to negative along the stream. The diffusion and dissipation term can be also decomposed into the diffusion term and the dissipation term, and the dissipation term is proved to be dominant.

Turbulent scalar flux

Turbulent scalar flux on the basis of the DNS data is evaluated and compared with the BML model (Libby and Bray 1981). Evolution of turbulent scalar flux, which is averaged between $t = 9.30$ ms and $t = 15.50$ ms, is shown in Fig. 6. In the BML model, the boundary of burned and unburned gases is defined as the progress variable of 0.5. It is shown that the counter-gradient diffusion dominates in the flame region, and that BML model well predicts it.

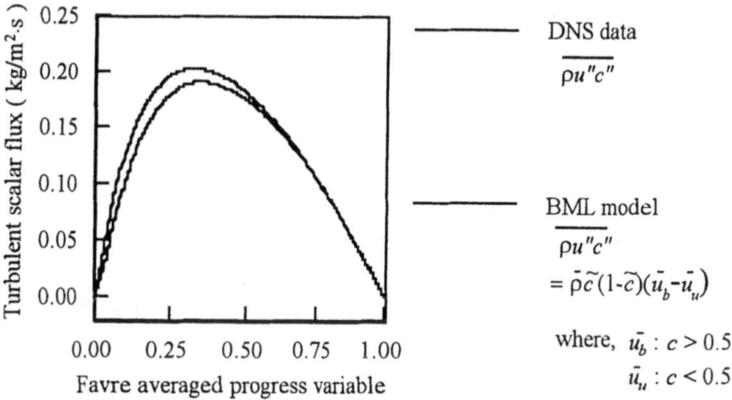

$$\overline{\rho u''c''}$$

$$\overline{\rho u''c''} = \bar{\rho}\tilde{c}(1-\tilde{c})(\bar{u}_b-\bar{u}_u)$$

where, $\bar{u}_b : c > 0.5$
$\bar{u}_u : c < 0.5$

Fig. 6. Evolution of turbulent scalar flux in the flame region.

The transport equation of turbulent scalar flux is written as follows:

$$\frac{\partial \overline{\rho u_j''c''}}{\partial t} + \frac{\partial}{\partial x_i}\left(\overline{\rho u_j''c''\tilde{u}_i}\right) = \underbrace{-\frac{\partial}{\partial x_i}\left(\overline{\rho u_i''u_j''c''}\right)}_{(A)} \underbrace{-\overline{\rho u_j''u_i''}\frac{\partial \tilde{c}}{\partial x_i}}_{(B)} \underbrace{-\overline{\rho c''u_i''}\frac{\partial \tilde{u}_j}{\partial x_i}}_{(C)}$$

$$\underbrace{-\overline{c''}\frac{\partial \bar{p}}{\partial x_j}}_{(D)} \underbrace{-\overline{c''\frac{\partial p'}{\partial x_j}}}_{(E)} \underbrace{+\frac{1}{Re}\frac{1}{Pr}\overline{u_j''\frac{\partial}{\partial x_i}\left(\frac{\partial c}{\partial x_i}\right)}}_{(F)} \underbrace{+\frac{1}{Re}\overline{c''\frac{\partial \tau_{ji}}{\partial x_i}}}_{(G)} \underbrace{+\overline{u_j''w}}_{(H)}$$

where term (A) is the transport by turbulence, term (B) is the source due to progress variable gradient, term (C) is the source due to velocity gradient, term (D) is the source due to pressure gradient, term (E) is the source due to fluctuating pressure, term (F) is the dissipation due to diffusivity, term (G) is the dissipation due to viscosity, term (H) is the velocity-reaction rate correlation.

Streamwise balance of the production rate of turbulent scalar flux is shown in Fig. 7. As shown in the figure, the mean pressure gradient term, the velocity-reaction rate correlation term and the fluctuating pressure term are positive sources to produce the counter-gradient diffusion. Mean velocity gradient term, mean progress variable gradient term and dissipation terms are negative sources.

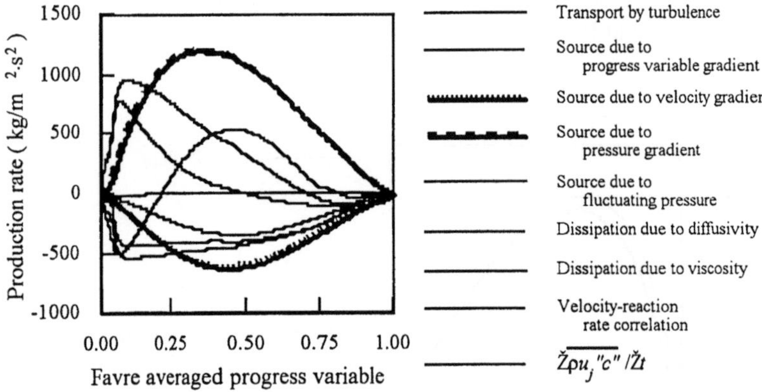

Fig. 7. Streamwise balance of the production rate of turbulent scalar flux as a function of mean progress variable. $\partial \overline{\rho u_j'' c''} / \partial t$ is calculated from the transport equation by subtracting the convection term from the right hand side.

Conclusions

A fully developed stationary wrinkled flame in a homogeneous turbulence flow is directly simulated with a single-step irreversible reaction. Following conclusions are obtained.

(1) Turbulent fluctuations, especially streamwise component, increase dramatically in the flame region, and an anisotropic turbulence is generated in the flame region.

(2) Pressure gradient term and pressure work term increase turbulent kinetic energy in the flame region, while diffusion and dissipation term and velocity gradient term decrease it. In the pressure work term, the pressure dilatation always has positive value, but the pressure diffusion changes its sign from

positive to negative. Dissipation is dominant in the diffusion and dissipation term.

(3) Counter-gradient diffusion dominates turbulent scalar flux and the BML model well predicts it.

(4) Mean pressure gradient term, velocity-reaction rate correlation term and fluctuating pressure term play important roles on production of counter-gradient diffusion, while mean velocity gradient term, mean progress variable gradient term and dissipation terms suppress it.

Acknowledgment

This work is partially supported by the Grant-in-Aid for Scientific Research by the Japan Society for the Promotion of Science (No. 2054).

References

Baum M, Poinsot T, Thévenin T (1994) Accurate Boundary Conditions for Multicomponent Reactive Flows. Journal of Computational Physics 116:247-261

Furukawa J, Noguchi Y, Hirano T (2000) Investigation of Flame Generated Turbulence in a Large-Scale and Low-Intensity Turbulent Premixed Flame with a 3-Element Electrostatic Probe and a 2-D LDV. Combustion Science and Technology 154:163-178

Hasegawa T, Morifuji T, Borghi R (1999) Direct Numerical Simulation of a Turbulent Premixed Flame. In: Proceedings of the 5th ASME/JSME Joint Thermal Engineering Conference AJTE99-6316

Kobayashi H, Nakashima T, Tamura T, Maruta K, Niioka T (1997) Turbulence Measurements and Observations of Turbulent Premixed Flames at Elevated Pressures up to 3.0 MPa. Combustion and Flame 108:104-117

Kobayashi H, Kawabata Y, Maruta K (1998) Experimental Study on General Correlation of Turbulent Burning Velocity at High Pressure. In: Twenty-Seventh Symposium (International) on Combustion. The Combustion Institute, pp 941-948

Libby P A, Bray K N C (1981) Countergradient Diffusion in Premixed Turbulent Flames. AIAA Journal 19:205-213

Nishiki S, Hasegawa T, Himeno R (2000) Numerical Study on Transport Properties of Turbulent Premixed Flames. In: Proceedings of the 3rd International Symposium on Turbulence, Heat and Mass Transfer, pp 815-822

Poinsot T J, Lele S K (1992) Boundary Conditions for Direct Simulations of Compressible Viscous Flows. Journal of Computational Physics 101:104-129

Rutland C J, Cant R S (1994) Turbulent Transport in Premixed Flames. In: Proceedings of the Summer Program. Center for Turbulence Research, NASA Ames/Stanford University, pp 75-94

Tanahashi M, Nada Y, Fujimura M, Miyauchi T (1999) Fine Scale Structure of H_2-Air Turbulent Premixed Flames. In: Proceedings of First International Symposium on Turbulence and Shear Flow Phenomena (in press)

Trouvé A, Poinsot T (1994) The Evolution Equation for the Flame Surface Density in

Turbulent Premixed Combustion. Journal of Fluid Mechanics 278:1-31

Flame Structure and Emission Characteristics of a Jet Stirred Reactor

Akira Yoshida[1], Hiroyoshi Naito[1], and Michinori Narisawa[2]

[1]Department of Mechanical Engineering, Tokyo Denki University, Tokyo 101-8457, Japan

[2]Research Institute, Ishikawajima Harima Heavy Industries Co., Ltd., Tokyo 135-8732, Japan.

Summary. A jet stirred reactor was designed for the study of high-intensity combustion using highly preheated mixture of methane and air. NO_X emission characteristics were measured experimentally. These measurements revealed that the high intensity and lean combustion with extremely low NO_X emission could be achieved. Schlieren photographs were taken to visualize the flame structure. Also the mean temperature, species concentration and ionization measurements were made. Schlieren photograph and mean temperature showed the uniformity of the temperature field, whereas species concentration and ion current showed that the main reaction zone is localized at the interface of high speed reacting mixture jet and recirculating combustion gas. However, the thickness of the mean reaction zone is thicker than that of the laminar flamelets. The assumption of the thin laminar flamelets is no more valid and the volumetric combustion must be achieved.

Key words. Thermal NO, Prompt NO, Laminar Flamelet, Stirred Reactor, Distributed Reaction Zone

Introduction

Recently, highly preheated air combustion is found paradoxically to have the potential to reduce the NO_X emission (Katsuki 1998), because highly preheated air makes it possible to extend the inflammability limit, and to maintain the flame even with the extremely low fuel concentration, which leads to the low NO_X emission. In the industrial burner, highly preheated air, which is provided by the heat regenerator, has been used. By the highly preheated air, the high intensity combustion can be achieved with low NO_X formation. However, the details of this combustion mechanism are not known well yet.

From viewpoint of the fundamental aspect of the flame structure, this type of flame is very interesting and attractive. From the early stage of the turbulent combustion studies, it has been said that the strong turbulence that can realize the high intensity combustion destroys the laminar flame structure and that the distributed reaction zone is achieved by the well-stirred reactor (Longwell and Weiss 1955, Malte and Pratt 1974). However, the existence of the distributed reaction zone has not been established still now. When the mixture temperature is low, the wrinkled laminar flame can be easily extinguished. Considering these backgrounds, it is interesting to know whether the distributed reaction zone is achieved by the highly preheated air combustion.

In the present study, the air is preheated by the regenerator up to 1400 K and mixed with methane just before entering the cylindrical jet stirred reactor. The mixture is preheated to the higher temperature than the auto-ignition temperature and injected into the jet stirred reactor with very high-speed. The injected mixture impinges the top wall of the reactor and recirculates within the reactor and the chemical reaction completes before it goes out of the reactor through the exhaust ports. The flame structure and the emission characteristics were measured by several methodologies.

Experimental apparatus

Figure 1 shows the jet stirred reactor and mixture supply system. The inner diameter of the reactor is 30 mm and the depth is 10 mm. Both sides are fitted with the quartz windows for visual and photographic observations. Highly preheated air provided by the regenerator is supplied from the bottom tube that is made of Inconel. Methane is mixed with the high temperature air in the small mixing chamber and immediately after mixing, mixture is injected with high-speed into the reactor before auto-ignition occurs. The mixture injected into the reactor begins to react just after injection, and impinges the top wall of the cylindrical reactor. Then reacting gas recirculates on left and right sides of the reactor. The combustion gas is exhausted from ports located at the bottom wall.

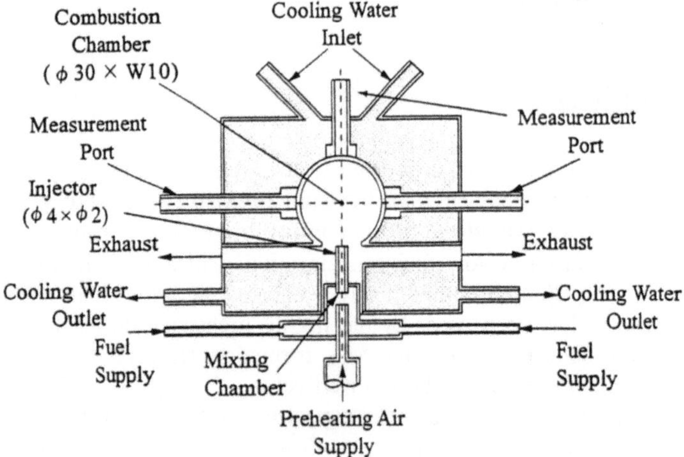

Fig. 1.   Jet Stirred Reactor.

As described later, the complete combustion occurs during the very short residence time. Three ports are located for measurements, two on the sidewall on the centerline and one on the top wall. The reactor can be water-cooled. However, the combustion characteristics are influenced by the wall temperature. Therefore, in the present study, the cooling water is not used.

For the temperature measurements, Pt/Pt-Rh13% thermocouple of the wire diameter of 0.3 mm was used. Gas samples were aspirated by the water-cooled stainless steel tube of the inner diameter of 0.8 mm and NO_X was analyzed by two chemiluminescence NO_X analyzers and CO, CO_2 and O_2 by respective testers. Ionization measurements were made using a water-cooled platinum wire of 0.1-mm diameter and 0.2 mm in length, and the stirred reactor was biased by 18 V. Stable species ωερε analyzed by the gas chromatograph.

Fig. 2.   Schlieren Photograph, $\phi = 1.0$, $T_j = 953$ K and $V_j = 264$ m/s.

Schlieren photographs were taken by the spark light source of which exposure time is 10 nanoseconds, which is short enough to freeze the high-speed phenomena in the reactor on the film in standard 35-mm camera.

Results and discussion

Schlieren photograph

Figure 2 show the schlieren photograph taken under the condition of equivalence ratio. $\phi = 1.0$, mixture temperature $T_J = 953$ K, and injection velocity $V_J = 264$ m/s. In this photograph, we cannot see any clear schlieren images, which mean that there is not any steep density gradient in the burner. This fact suggests that the temperature in the reactor is uniform. Therefore, this flame structure is different from that of the so-called wrinkled laminar flame stabilized on the Bunsen burner.

Temperature and emission characteristics of jet stirred reactor

As described above, we found that there exists no clear density gradient qualitatively by the schlieren photographs. We further examined the uniformity of the scalar quantities; the distributions of mean temperature, NO_X and other species were measured. These are shown in Figures 3 (a) and (b) under the conditions of $\phi = 1.0$, $T_J = 959$ K and $V_J = 266$ m/s for (a) and $\phi = 1.0$, $T_J = 1181$ K and $V_J = 329$ m/s for (b). Except for the center region and the wall boundary, distributions of mean temperature and species concentration are nearly flat for both cases. Around the centerline, the O_2 concentration is high and then decreases with the radial distance and the CO_2 concentration increases on the contrary. At the radial

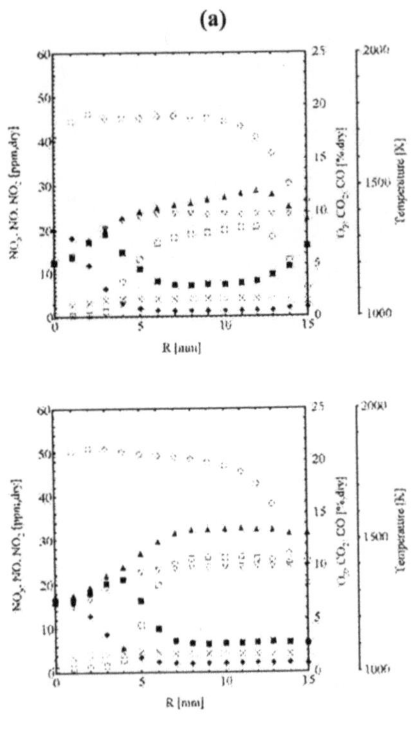

(a)

(b)

Fig. 3. Temperature and Species Concentration Distributions.

    (a) $\phi = 1.0$, $T_j = 959$ K and $V_j = 266$ m/s.

    (b) $\phi = 1.0$, $T_j = 1180$ K and $V_j = 329$ m/s.

$\bigcirc$;Temperature, $\blacktriangle$;NO_x, $\triangledown$;CO_2, $\square$;NO, $\blacksquare$;NO_2, $\times$;CO, $\blacklozenge$;O_2.

distance R = 6 mm, they become constant at their equilibrium values. This fact suggests that the main reaction occurs in the mixing region of the jet and recircurating gas and that although the residence time is extremely short, the mixture burns out completely in the reactor. And the mean reaction zone thickness increases as compared to the laminar flame.

Figure 3 (a) shows a slight temperature decrease on the centerline. This fact suggests that the chemical reaction is proceeding in this region as shown by the distribution of the species concentrations. For Fig. 3 (b), although the concentration distribution remains, the mean temperature distribution becomes constant within the reactor that shows that the turbulent mixing is very fast.

The excess enthalpy is added to the mixture and the reaction temperature increases. Although the contribution of the NO_X originating from the Zel'dovich mechanism is small due to the short residence time, noticeable effect remains in the increase of NO_X due to the higher gas temperature.

Figure 4 shows the dependence of the NO_X formation at the exhaust port on the temperature and the equivalence ratio. Here we must consider the three NOx

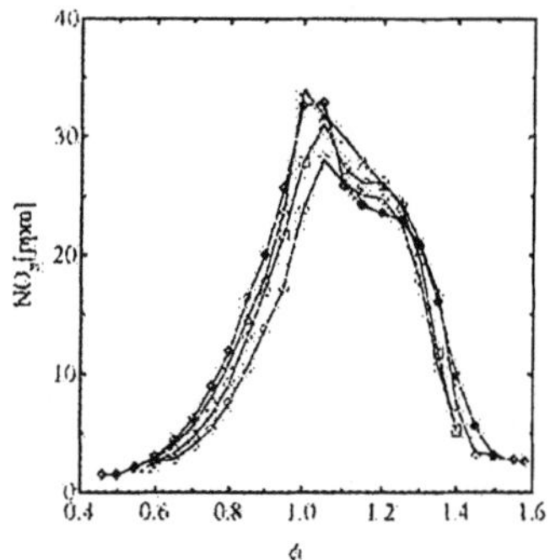

Fig. 4. Dependence of NO_x Concentration at the Exhaust Port on Temperature and Equivalence Ratio. $\bigcirc$;T_j = 873 K, $\square$;T_j = 973 K, $\triangle$;T_j = 1073 K, $\diamondsuit$;T_j = 1173 K.

production mechanisms, the Zel'dovich mechanisms (Katsuki 1998), the prompt NO mechanism introduced by Fenimore (1971) and the nitrous oxide mechanism suggested by Malte and Pratt (1974). All three mechanisms can contribute in the case of lean and rich premixed flame. The nitrous oxide was developed in order to explain relatively large amount of NO_X measured from a CO-fired jet stirred reactor. Therefore, the nitrous oxide mechanism need not take into consideration here. Figure 4 shows a little dependency on the preheating temperature (873 K – 1173 K). This weak dependency is due to a little contribution of the Zel'dovich mechanism, which depends strongly on the temperature. The remaining is the prompt NO mechanism. On the rich side, NO_X decreases rather gradually. This is the typical feature of the prompt NO. It should be worth saying that ϕ = 0.45, the NO_X concentration is as low as 1.4 ppm. The emission index for this case was 0.033. With high T_j, the stable combustion is realized at oxygen concentration lower than the limit of inflammability at ambient temperature.

Therefore, high intensity and lean combustion was achieved with low NO_X

emission by the jet stirred reactor using highly preheated air. Steele et al. (1995) made the same measurements in the well-stirred reactor with preheated methane-air mixture. The experimental condition and the results are as follows; for T_J = 583 K, ϕ = 0.542, the reactor temperature T_R = 1589 K and the residence time τ = 6.28 ms, the NO_X concentration is 3.2 ppm. T_J is about 400 K and T_R is about 300 K lower than those of the present experiment. However, the NO_X concentration measured in the present study under similar conditions is about half of that measured by Steele et al. (1995). The reason why so low NO_X emission is realized should be the difference of the residence time. Therefore, the prompt NO mechanism may be dominant in the present study. As shown in Fig. 4, the peak value of the NO_X appears at ϕ = 1.05 where the reactor temperature is the highest. Therefore, the Zel'dovich mechanism is not negligible. The residence time will increase because a part of the reacting spots and the combustion product should recirculate in the reactor. However, even in the range of ϕ = 1.2 to 1.3, the NO_X emission level is quite high. This fact suggests that the characteristic features of the prompt NO are predominant. On the rich side, the NO_X originating from the Zel'dovich mechanism should be suppressed and NO_X originating from the prompt NO mechanism may increase. The NO_X emission level is extremely lower than that produced by the opposed jet burner (Yoshida et al. 1997).

Flame surface density

As shown by Fig. 3, the mean reaction zone is thick. If we assume that the thick reaction zone consists of the fluctuating thin laminar flamelets, which should be tangled or multiply folded, the flame surface area can be estimated. The laminar burning velocity of methane at high temperature is calculated by the empirical equation obtained by Andrews and Bradley (1972). The resulting surface area is surprisingly small, 167 mm^2 for V_J = 262 m/s and T_J = 953 K and 335 mm^2 for V_J = 525 m/s and T_J = 953 K. The global flame surface density $\langle\Sigma\rangle$ are 2.18 x $10^{-2} mm^{-1}$ and 4.74 x 10^{-2} mm^{-1}, respectively. If the laminar flame zone can be assumed, the value of $\langle\Sigma\rangle$ is extremely small (Cant et al. 1990, Veynante et al. 1996). In other words, the flame area is 21.8 μm^2 and 47.4 μm^2 per cubic mm, respectively. And combustion reaction completed at the exhaust port. Such small flame areas are non-realistic from the wrinkled laminar flamelet concept, even though the preheat zone may be intruded by the small eddies (Buschmann et al. 1996, Dinkelacher et al. 1998). This fact suggest that the assumption of the flamelets is no more valid. Instead, discrete reaction spots are embedded in the burned or unburned gases and volumetric reaction occurs.

Ionization measurement

In the present study, local ionization measurements were made to obtain the characteristic features of the reaction zone structure in the jet stirred reacer.

Figure 5 shows the typical ionization signal for T_j =953 K, V_j = 262 m/s and ϕ = 1.0. Figures (a), (b) and (c) correspond to the signals at R = 0 mm, 3.75 mm and

7.5 mm, respectively. At R = 0 mm, the signal form is random. And paired spikes which are the characteristic feature of the wrinkled laminar flame is unclear. Even when the mixture enters the jet stirred reactor, the reaction has started, because T_j is higher than the auto-ignition temperature. It is apparent that the gas spots of intermediate reaction level intermingle with each other and the distributed reaction zone will be realized. At R = 3.75 mm, where the oxygen concentration decreases and CO_2 concentration increases. At the same time, the jitter of the ionization increases. This fact suggest that the reaction in spots proceeds.

At R = 7.5 mm, where both O_2 and CO_2 concentrations become constant at their equilibrium values, the mean ion current becomes low and the chemical reaction approaches toward completion in all the reaction spots. The signal shape becomes peaky and random. With increase of V_j to 525 m/s, typical ionization signal changes as shown in Fig. 6. The mean and fluctuating values of the ion current on the center line, Fig. 5 (a), become higher than those of V_j = 262 m/s. On the centerline, reaction spots are highly ionized and the jitter also appeared. With the increase of V_j, the turbulence intensity also increases. Therefore, the jitter becomes large. The reaction in the reacting spots become to completion.

Although the time during which the mixture element reaches from the mixing

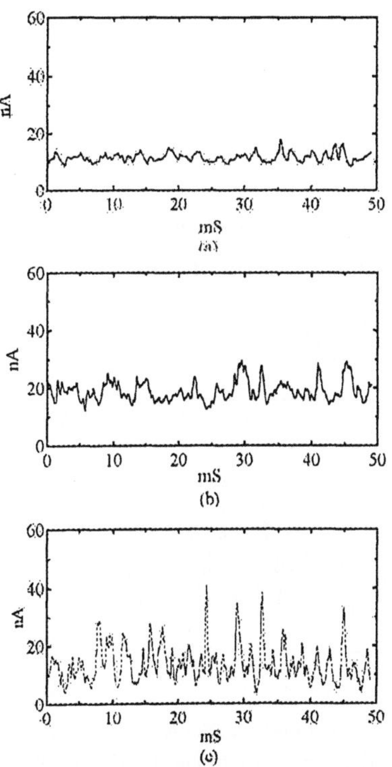

Fig. 5. Typical Ionization Signal for T_j = 953 K, V_j = 262 m/s and ϕ = 1.0.

chamber to the measuring point is shorter for Fig. 6 than that for Fig. 5, chemical reaction of the latter case proceeds faster than the former. This is due to the fact that the turbulence intensity is higher and the small-scale eddies break the laminar flame sheet. Therefore, reaction rate becomes lower than the wrinkled laminar flame, because the reaction zone diffuses throughout the volume. At R = 3.75 mm, Fig. 5 (b), which is the outer edge of the oxygen reduction zone, the fluctuation of the ion current increases and the mean ion current decreases as compared to that of the centerline. Therefore, though in some spots, chemical reaction is still proceeding, the number of the reacting spots in which the chemical reaction approaches toward completion increases. At R = 7.5 mm, Fig. 5 (c) where both O_2 and CO_2 concentrations become constant at their equilibrium values, the mean ion current decreases further and chemical reaction tends to completion.

With the increase of R, mean ionization decreases. On the contrary, amplitude of jitter increases and many random spikes appears. This is the characteristics of the distributed reaction zone. Therefore, with the increase of the mixture velocity, reaction rate is increased by the strong turbulence, and features of the wrinkled laminar flamelets disappear even on the centerline. The integral time scales were

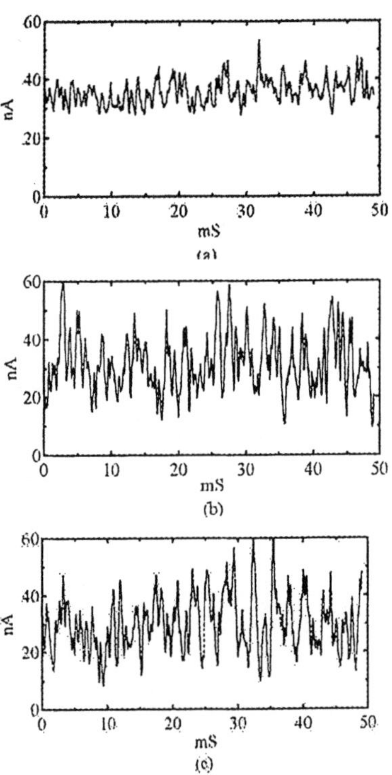

Fig. 6. Typical Ionization Signal for T_j = 953 K, V_j = 525 m/s and ϕ = 1.0.
(a) R = 0 mm, **(b)** R = 3.75 mm and **(c)** R = 7.5 mm.

measured from the auto-correlation curve for cold flow. Figure 7 shows the integral time scale for $\phi = 1.0$ and $V_j = 241$ m/s to 324 m/sec. The data scattered,

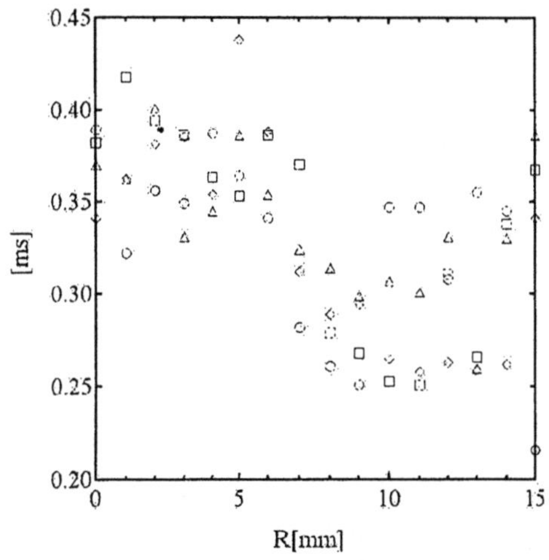

Fig. 7. Integral Time Scale.
○; Tj = 873 K, □; Tj = 973 K, △; Tj = 1073 K, ▽; 1173 K.

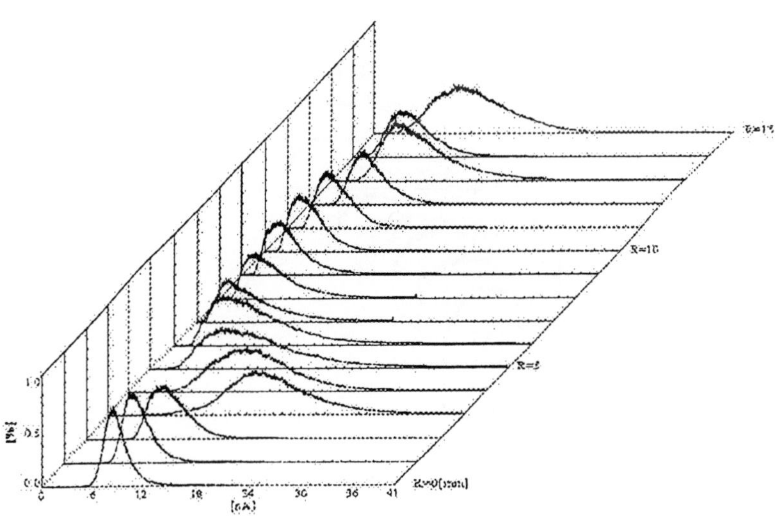

Fig. 8.   Probability Density Function of the Ion Current.

however, the integral time scales on the unburned gas side is apparently larger than those of the burned gas side. Although we did not measured the velocity field with combustion, the velocity should be high around the centerline. Therefore, the integral scale of the reacting gas around the centerline is larger than the outer recirculation zone which consists of the burned spots. The ionization signal form is extremely different from that of wrinkled laminar flame of which characteristics are paired spikes (Suzuki et al. 1978).

Figure 8 shows probability density functions of the ion current for T_j = 1173 K and V_j = 324 m/s. On the centerline, the PDF profile is nearly Gaussian, and the measured value is finite. This will be attributed to the fact that the chemical reaction occurs, because the temperature of the mixture is above the auto-ignition temperature. However the reaction is not so strong. In the main reaction zone, the maximum ion current increases, although the PDF of high ionization is low. In this region (R = 3 mm to 6 mm), main chemical reaction occurs, because the probability of high ion concentration appears. In the further outer region, weak ionization is observed and sometimes it goes to zero. Therefore, in this region, the reaction of the reacting spots goes to completion.

Figure 9 shows the power spectra of the ionization signal for Tj = 1173 K and Vj = 324 m/s. It is surprising that the peak appears below 1 kHz. For reference,

from the cold flow experiments (V_j = 163 m/s), it was found that the integral scale is about 30 mm on the centerline and decrease to about 5 mm in the main reaction zone and outer zone, whereas the Kolmogorov microscale is nearly constant at 40

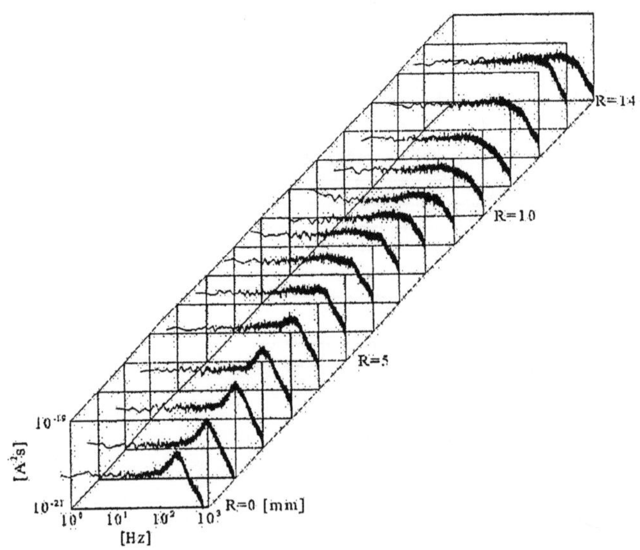

Fig. 9. Power Spectra of the Ion Current Signal.

μm. Therefore, the scale of the reacting spots   in the combustion field should be large. Although power spectra in the main reaction zone increases a little, peak frequency decreases gradually on the burned gas region.

In the jet stirred reactor, distributed reaction zone has been observed.. The reaction spots soon coalesce with each other to make the large reacting spots in the outer region where turbulence decays.

Conclusions

The jet stirred reactor was designed for the study of high intensity combustion using highly preheated mixture of methane and air. The NO_X emission characteristics were measured and flame structure was discussed from the flame surface density and the ionization measurements. Conclusions of this study are summarized as follows.

Schlieren photographs show that there is not any density gradient in the jet stirred reactor in which the highly preheated mixture is burnt. The temperature distributions are nearly constant in the jet stirred reactor. This fact is attributed to the fast mixing due to strong turbulence.

The high intensity and lean combustion with extremely low NO_X emission can be achieved in the jet stirred reactor by using the highly preheated air.

In the jet stirred reactor, main chemical reaction is localized around the mixture jet. Temperature distribution is flat, whereas species distribution and ion current are uniform except for the main chemical reaction zone.

The distributed reaction zone is achieved by the jet stirred reactor, that is confirmed by the flame surface density and ion current measurements.

By using the highly preheated mixture, extremely lean combustion can be achieved., and extremely low NO_x combustion is realized.

Wave form of the ion current of the distributed reaction zone is random and different from that of the wrinkled laminar flame of which feature is paired spikes.

References

Andrews GE, Bradley D (1972) The burning velocity of methane-air mixtures. Combust Flame 19:275-288

Buschmann A, Dinkelacker F, Schafer T, Schafer M, Wolfrum J (1996) Measurement of the instantaneous detailed flame structure in turbulent premixed combustion. Twenty-Sixth Symposium (International) on Combustion, The Combustion Institute, PA., pp. 437-445

Cant RS, Pope SB, Bray KNC (1990) Modelling of flamelet surface-to-volume ratio in turbulent premixed combustion. Twenty-Third Symposium (International) on Combustion, The Combustion Institute, Pittsburgh, PA, pp.809-815

Dinkelacker F, Soika A, Most D, Hofmann D, Leipertz A, Polifke W (1998) Structure of locally quenched highly turbulent lean premixed flames. Twenty-Seventh Symposium (International) on Combustion, The Combustion Institute, Pittsburgh, PA., pp. 857-865

Fenimore CP (1971) Formation of nitric oxide in premixed hydrocarbon flame. Thirteenth

Symposium (International) on Combustion, The Combustion Institute, Pittsburgh, PA, pp.373-380

Katsuki M, Hasegawa T (1998) The science and technology of combustion in highly preheated air. Twenty-Seventh Symposium (International) on Combustion, The Combustion Institute, Pittsburgh, PA, pp. 3135-3146

Longwell JP, Weiss MA (1955) High temperature reaction rates in hydrocarbon combustion. Ind. Eng. Chem. **47**:1634-1643

Malte PC, Pratt DT (1974) Measurement of atomic oxygen and nitrogen. Fifteenth Symposium (International) on Combustion, The Combustion Institute, Pittsburgh, PA, pp. 1061-1070

Suzuki T, Hirano T, Tsuji H (1978) Flame front movements of turbulent premixed flame. Seventeenth Symposium (International) on Combustion, The Combustion Institute, Pittsburgh, PA., pp. 289-297

Steele RC, Malte PC, Nicol DG, Kramlich JC (1995) NO_x and N_2O in Lean-Premixed Jet-Stirred Flames. *Combust. Flame* **100**:440-449 (1995).

Veynante D, Piano J, Duclos JM and Martel C (1996) Experimental analysis of flame surface density models for premixed turbulent combustion. Twenty-Sixth Symposium (International) on Combustion, The Combustion Institute, Pittsburgh, PA., pp. 413-420

Yoshida A, Hirose, K, Kotani Y (1997) Structure and NO_x formation in premixed and non-premixed opposed jet burner. Proc. 4th APISCEU, vol. 1, pp. 304-309

The Unstable Motion of Cellular Premixed Flames Caused by Intrinsic Instability

Satoshi Kadowaki

Department of Mechanical Engineering, Nagoya Institute of Technology
Gokiso-cho, Showa-ku, Nagoya 466-8555, Japan
Department of Mechanical Engineering, Nagaoka University of Technology
Kamitomioka, Nagaoka 940-2188, Japan

Summary. The unsteady calculations of two-dimensional (2-D) and three-dimensional (3-D) reactive flows are performed to study the unstable motion of cellular premixed flames caused by intrinsic instability. The compressible Navier-Stokes (N-S) equation including a one-step irreversible chemical reaction is employed. An infinitesimal disturbance is superimposed on a stationary planar flame to obtain the relation between the growth rate and the wave number, i.e., the dispersion relation. As the Lewis number becomes lower, the growth rate increases and the unstable range widens. For upward propagating flames, the growth rate increases and the unstable range widens as the acceleration increases. In addition, the dispersion relation of 2-D flames is nearly the same as that of 3-D flames. The disturbance with the peculiar wavelength corresponding to the maximum growth rate is superimposed to study the unstable motion of cellular flames. When the Lewis number is lower than unity, the lateral movement of cells is observed. The lateral velocity of cellular flames increases as the Lewis number becomes lower. In addition, the lateral velocity of 3-D flames is about twice that of 2-D flames. The body-force effect has an influence on the lateral velocity. For upward propagating flames, the lateral velocity decreases as the acceleration increases, even though intrinsic instability becomes stronger. The reason is that the high-temperature region behind a convex flame front with respect to the unburned gas becomes larger as the acceleration increases.

Key words. Unstable Motion, Cellular Flame, Intrinsic Instability

Introduction

We have three basic effects responsible for the intrinsic instability of premixed flames, i.e., the hydrodynamic, diffusive-thermal, and body-force effects (Williams 1985). Most of cellular flames are formed owing to intrinsic instability, and stable or unstable motion is observed in experiments (Sabathier et al. 1981, Searby and Quinard 1990, Pearlman and Ronney 1994). There are a large number of studies on the flame instability, and they are reviewed by Sivashinsky (1983, 1990) and Clavin (1985).

The hydrodynamic effect is generated by the thermal expansion of reactive gases and has a destabilizing influence. This type of instability is called Darrieus-Landau instability. The hydrodynamic effect is essential to the intrinsic instability of premixed flames, since reactive gases expand thermally when they pass through the flame front.

The diffusive-thermal effect is due to the interaction of diffusion and heat-conduction processes. This effect has a destabilizing influence at Lewis numbers of deficient reactant lower than unity and has a stabilizing influence at high Lewis numbers. Thus, cellular flames are more likely to be observed at low Lewis numbers. The diffusive-thermal effect is studied by Barenblatt et al. (1962), Sivashinsky (1977), and Joulin and Mitani (1981).

The body-force effect is due to the difference in density between above and below fluids. This effect has a destabilizing influence when the density of an above fluid is larger than that of a below fluid, which is called Rayleigh-Taylor instability. In premixed flames, the density is drastically changed through the flame front. Thus, the body-force effect has a destabilizing influence on upward propagating flames and has a stabilizing influence on downward propagating flames. The body-force effect on the flame instability is studied by Markstein (1964), Matkowsky and Sivashinsky (1979), and Pelce and Clavin (1982).

On the unstable motion of cellular premixed flames, numerical study based on the diffusive-thermal model equation is performed by Michelson and Sivashinsky (1982), Margolis and Matkowsky (1983), and Cambray and Joulin (1994). Unstable motion appears at sufficiently low Lewis numbers. In the diffusive-thermal model equation, the constant-density approximation is used, so that the results are valid only for flames with small heat release. After that, numerical study based on the compressible N-S equation, where the hydrodynamic effect is taken into account, is performed by Patnaik and Kailasanath (1994), Denet and Haldenwang (1995), Bychkov et al. (1996), and Kadowaki (1997, 1999). The hydrodynamic effect plays an important role, so that we need to employ the compressible N-S equation to study the unstable motion of cellular flame fronts.

In this paper, we arrange our former studies and discuss the unstable motion of 2-D and 3-D cellular premixed flames caused by intrinsic instability. The compressible N-S equation including a one-step irreversible chemical reaction is employed, and unsteady reactive flows are calculated. We superimpose an infinitesimal disturbance on a stationary planar flame to obtain the relation between the growth rate and the wave number, i.e., the dispersion relation. The peculiar wavelength, which corresponds to the maximum growth rate, is determined by the dispersion relation. To study the unstable motion of cellular flames, the disturbance with the peculiar wavelength is superimposed, since the spacing between cells is equal to the peculiar wavelength (Kadowaki 1996). We calculate the unsteady behavior of cellular flame fronts to obtain the lateral velocity of cells, and examine the influence of diffusive-thermal and body-force instabilities on it.

Numerical procedure

The 2-D and 3-D unsteady reactive flows are treated in the present calculation. The chemical reaction is one-step irreversible, and the unburned and burned gases have the same molecular weights and the same Lewis numbers. Using Cartesian coordinates, we take the flame surface as the yz-surface, with the gas velocity in the positive x-direction. The compressible N-S equation written in the conservation form is employed. The flow variables are nondimensionalized by the preheat zone thickness, the burning velocity, and the density of the unburned gas. The preheat zone thickness is defined as the thermal diffusivity divided by the burning velocity.

We consider the premixed flame whose burning velocity is 0.83 m/s and adiabatic flame temperature is 2086 K at atmospheric pressure and room temperature. The burning velocity is sufficiently small compared with the sound velocity, so that the nondimensionalized results change very little, even though the burning velocity is changed. To study the influence of diffusive-thermal and body-force instabilities, we take the Lewis number $Le \leq 1$ and the acceleration G ≥ 0. Positive acceleration corresponds to upward propagating flames.

The explicit MacCormack scheme, which has second-order accuracy in both time and space, is adopted. Computational domain is 160 times the preheat zone thickness in the x-direction and one wavelength of a disturbance in y- and z-directions. This domain is resolved by a 341×65 variably spaced grid for 2-D flow and by a 261×31×53 variably spaced grid for 3-D flow. The minimum grid size in the x-direction is one-fifth of the preheat zone thickness. The grid is fine enough to prevent numerical errors from contaminating the solutions. We use free-flow conditions at boundaries in the x-direction, and spatially periodic conditions in y- and z-directions. The inlet-flow velocity is set to the burning velocity.

The calculation starts from the initial state provided by the solution of a stationary planar flame, on which we superimpose a disturbance periodic in y- and z-directions. The displacement of the flame front in the x-direction due to the disturbance for 2-D flames is given by $\sin(2\pi y/\lambda_2)$, where λ is the wavelength. For 3-D flames, a hexagonal disturbance is superimposed (Christopherson 1940). The displacement is given by $\sin(2\pi y/\lambda_y)\sin(2\pi z/\lambda_z) - \cos(4\pi z/\lambda_z)/2$, where the relation $\lambda_z/\lambda_y = \sqrt{3}$ is realized.

Dispersion relation

An infinitesimal disturbance is superimposed on a stationary planar flame to obtain the relation between the growth rate and the wave number, i.e., the dispersion relation. The initial amplitude of a disturbance is set to 0.08, which is small enough to study the flame instability.

The evolution of a disturbed flame front of the 2-D flame for Le = 1.0, G = 0.0, and λ_2 = 102.4 is illustrated in Fig. 1. The location of the flame front is defined as the position where the reaction rate takes a maximum value. The abscissa is normalized by the initial amplitude A_0. The unburned gas flows in from the left, and the burned gas flows out to the right. The disturbance grows with time, maintaining a sinusoidal shape. The amplitude growth rate is shown in Fig. 2. The ordinate is the natural logarithm of the ratio of the amplitude of a disturbance to the initial amplitude. The amplitude growth rate grows linearly with time. Thus, the amplitude grows exponentially with time, which is consistent with the results of linear analyses. This behavior of a disturbance, growing exponentially with time, is observed only when a disturbance is sufficiently small. When the amplitude grows to some degree, the growth rate is gradually lowered and eventually drops to zero.

Changing the wavelength of a disturbance, we obtain the relation between the growth rate ω and the wave number k. The dispersion relations of 2-D and 3-D flames for Le = 0.5, 0.7, and 1.0, and G = 0.0 are shown in Fig. 3. Closed and open symbols denote 2-D and 3-D flames, respectively. The growth rate is positive (negative) at small (large) wave numbers. The wave number separating the stable/unstable range is the marginal wave number. As the Lewis number becomes lower, the growth rate increases and the unstable range widens, which is due to the diffusive-thermal effect. In addition, the dispersion relation of 2-D flames is nearly the same as that of 3-D flames. Figure 4 shows the dispersion relations of 2-D upward propagating flames for Le = 1.0 and G = 0.0, 0.5, 1.0, and 1.5. As the acceleration increases, the growth rate increases and the unstable range widens, since the body-force effect contributes to the destabilization of flame fronts.

Cellular flame

The disturbance with the peculiar wavelength is superimposed to study the unstable motion of cellular flames. The front shapes of 2-D flames for Le = 0.5, 0.6, and 0.8, and G = 0.0 are illustrated in Fig. 5. The unburned gas flows in from the left at the burning velocity. The disturbance grows initially with time, and the cellular flame is formed. The cellular flame moves upstream, indicating that the flame velocity is increased. After cellular flame formation, cells move in the y-direction, i.e., move laterally. The lateral velocity increases as the Lewis number becomes lower, since diffusive-thermal instability becomes stronger. The lateral movement of cells is observed at Le < 1. When the Lewis number is unity, no lateral movement appears.

To study the mechanism of the lateral movement of cells, the temperature distribution of the 2-D cellular flame for Le = 0.5 is illustrated in Fig. 6. The temperature has an overshoot at a convex flame front with respect to the unburned gas. This overshoot is not observed when the Lewis number is unity. The overshoot of temperature is generated by the diffusive-thermal effect and by the nonlinear effect of the flame front. In general, the physical phenomena associated with an overshoot are unstable. Thus, the overshoot of temperature causes a breaking of the reflection symmetry of cells, and then the lateral movement appears.

We treat 3-D flames. The front shapes of the cellular flame for Le = 0.5, G = 0.0, λ_y = 13.3, and λ_z = 23.0 at t = 3 and 8 are illustrated in Fig. 7. The schematic domain is $3\lambda_y \times 2\lambda_z$. The unburned gas flows in from the bottom, and the burned gas flows out to the top. The cellular flame is formed at t = 3, and then moves laterally, just as the 2-D flame did. Figure 8 shows the distributions of cells in the yz-plane. Regularly arranged hexagonal cells are observed, and the spacing between cells is equal to λ_y. We find that cells move to the right-bottom of the figure, i.e., move laterally.

The lateral velocities of 2-D and 3-D cellular flames depending on the Lewis number are shown in Fig. 9. As the Lewis number becomes lower, the lateral velocity increases, since the instability level becomes higher. In addition, the lateral velocity of 3-D cellular flames is about twice that of 2-D cellular flames. The reason is that the maximum flame temperature at a convex flame front of 3-D cellular flames is higher than that of 2-D cellular flames.

Next, we study the body-force effect on the lateral movement of cells at low Lewis numbers. Figure 10 shows the lateral velocities of 2-D cellular flames for Le = 0.5, 0.6, and 0.8, depending on the acceleration. As the acceleration increases, the lateral velocity decreases, even though the instability level becomes higher.

We study the mechanism of the decrease in lateral velocity with an increase in acceleration. The temperature distributions of 2-D cellular flames for Le = 0.5 and G = 0 and 4 at a convex flame front are illustrated in Fig. 11. The temperature has an overshoot near the flame front. The maximum flame temperature of the G = 4 flame (= 7.79) is slightly higher than that of the G = 0 flame (= 7.74). A slight difference in maximum flame temperature has a slight influence on the change of lateral velocity. We have a high-temperature region behind a convex flame front. The high-temperature region of the G = 4 flame (= 3.5) is larger than that of the G = 0 flame (= 2.2), since the cell depth of the former flame (= 6.6) is larger than that of the latter flame (= 5.6). The augmentation in high-temperature region weakens the unstable motion of cellular flames. Thus, the lateral velocity decreases as the acceleration increases.

References

Barenblatt GI, Zeldovich YB, Istratov AG (1962) On diffusive-thermal stability of a laminar flame. J Appl Mech Tech Phys 4: 21-26

Bychkov VV, Golberg SM, Liberman MA, Eriksson LE (1996) Propagation of curved stationary flames in tubes. Phys Rev E 54: 3713-3724

Cambray P, Joulin G (1994) Length-scales of wrinkling of weakly-forced, unstable premixed flames. Combust Sci Technol 97: 405-428

Christopherson DG (1940) Note on the vibration of membranes. Quarterly J Math 11: 63-65

Clavin P (1985) Dynamic behavior of premixed flame fronts in laminar and turbulent flows. Prog Energy Combust Sci 11: 1-59

Denet B, Haldenwang P (1995) A numerical study of premixed flames Darrieus-Landau instability. Combust Sci Technol 104: 143-167

Joulin G, Mitani T (1981) Linear stability analysis of two-reactant flames. Combust Flame 40: 235-246

Kadowaki S (1996) Numerical study on the instability of premixed plane flames in the three-dimensional field. Int J Heat Fluid Flow 17: 557-566

Kadowaki S (1997) Numerical study on lateral movements of cellular flames. Phys Rev E 56: 2966-2971

Kadowaki S (1999) The lateral movement of the three-dimensional cellular flame at low Lewis numbers. Int J Heat Fluid Flow 20: 649-656

Margolis SB, Matkowsky BJ (1983) Nonlinear stability and bifurcation in the transition from laminar to turbulent flame propagation. Combust Sci Technol 34: 45-77

Markstein GH (1964) Nonsteady Flame Propagation. Pergamon, Oxford, pp 15-74

Matkowsky BJ, Sivashinsky GI (1979) Acceleration effects on the stability of flame propagation. SIAM J Appl Math 37: 669-685

Michelson DM, Sivashinsky GI (1982) Thermal-expansion induced cellular flames. Combust Flame 48: 211-217

Patnaik G, Kailasanath K (1994) Numerical simulations of burner-stabilized hydrogen-air flames in microgravity. Combust Flame 99: 247-253

Pearlman HG, Ronney PD (1994) Near-limit behavior of high-Lewis number premixed flames in tubes at normal and low gravity. Phys Fluids 6: 4009-4018

Pelce P, Clavin P (1982) Influence of hydrodynamics and diffusion upon the stability limits of laminar premixed flames. J Fluid Mech 124: 219-237

Sabathier F, Boyer L, Clavin P (1981) Experimental study of a weak turbulent premixed flame. Prog Astronaut Aeronaut 76: 246-258

Searby G, Quinard J (1990) Direct and indirect measurements of Markstein numbers of premixed flames. Combust Flame 82: 298-311

Sivashinsky GI (1977) Diffusional-thermal theory of cellular flames. Combust Sci Technol 15: 137-146

Sivashinsky GI (1983) Instabilities, pattern formation, and turbulence in flames. Ann Rev Fluid Mech 15: 179-199

Sivashinsky GI (1990) On the intrinsic dynamics of premixed flames. Phil Trans R Soc London A 332: 135-148

Williams FA (1985) Combustion Theory, 2nd ed. Addison-Wesley, Reading, pp 341-365

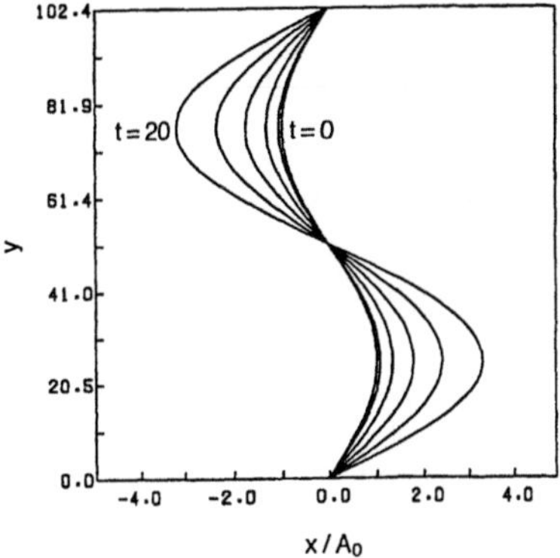

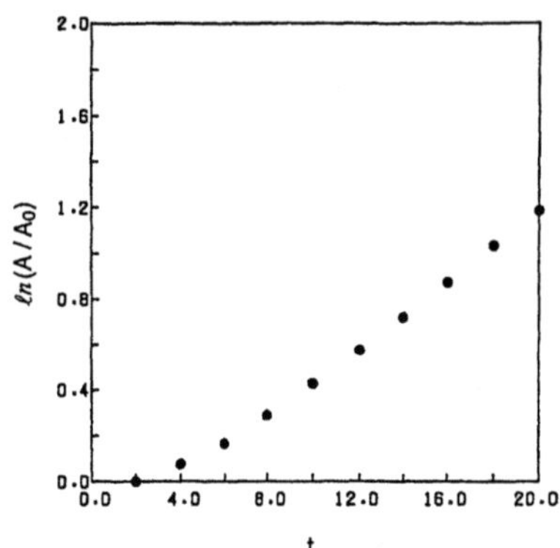

Fig. 1. Evolution of a disturbed flame front of the 2-D flame for Le = 1.0, G = 0.0, and λ_2 = 102.4 (t = 0, 4, 8, . . . , 20).

Fig. 2. Amplitude growth rate of the 2-D flame for Le = 1.0, G = 0.0, and λ_2 = 102.4 (t = 0 - 20).

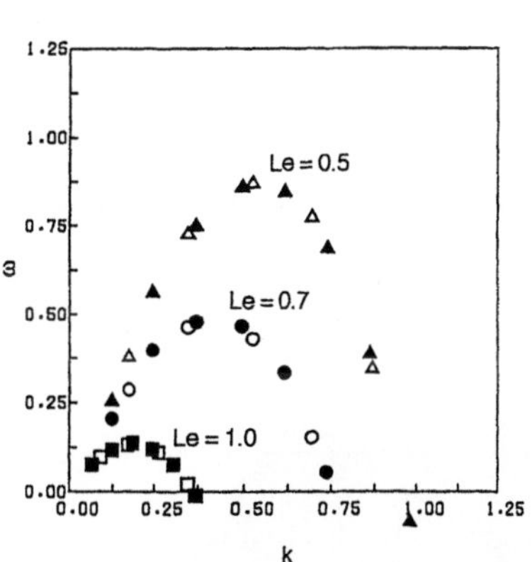

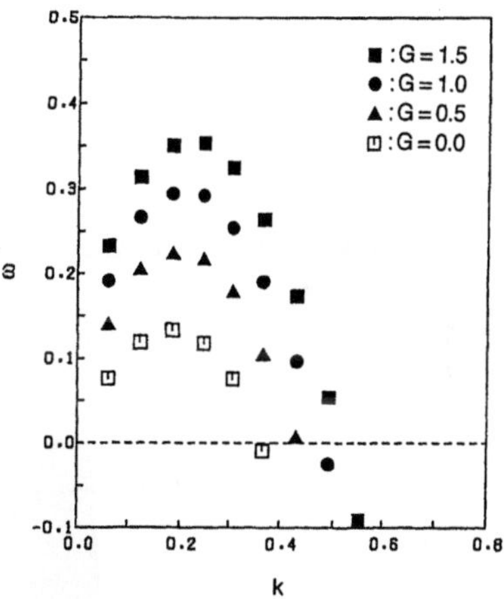

Fig. 3. Dispersion relations of 2-D and 3-D flames for Le = 0.5, 0.7, and 1.0, and G = 0.0; closed and open symbols denote 2-D and 3-D flames, respectively.

Fig. 4. Dispersion relations of 2-D upward propagating flames for Le = 1.0 and G = 0.0, 0.5, 1.0, and 1.5.

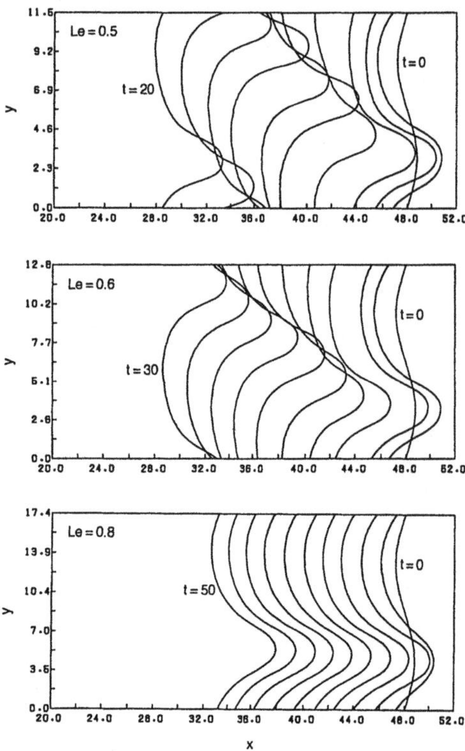

Fig. 5. Front shapes of 2-D flames for Le = 0.5, 0.6, and 0.8, and G = 0.0.

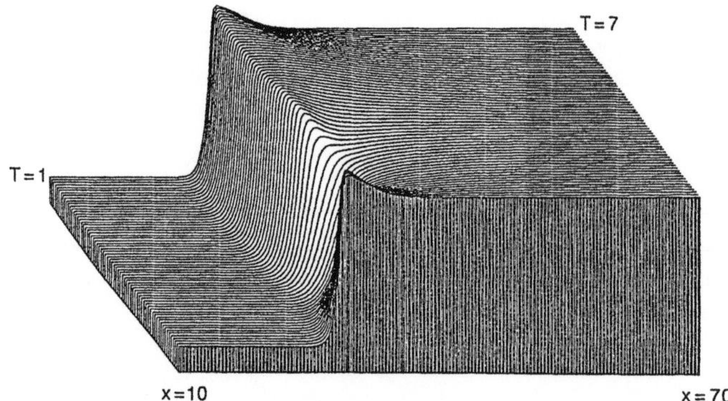

Fig. 6. Temperature distribution of the 2-D cellular flame for Le = 0.5 and G = 0.0 (t = 20)

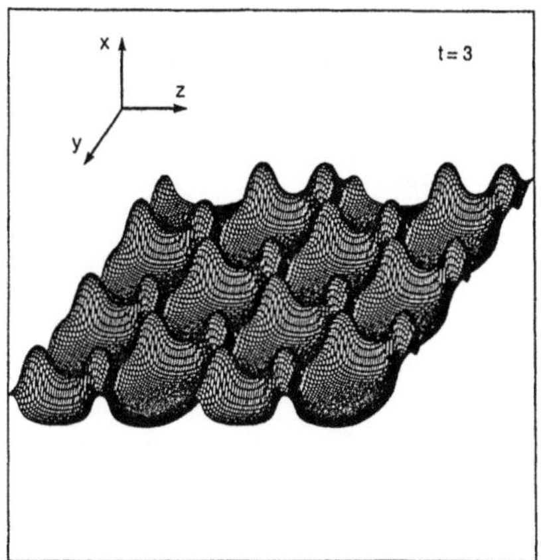
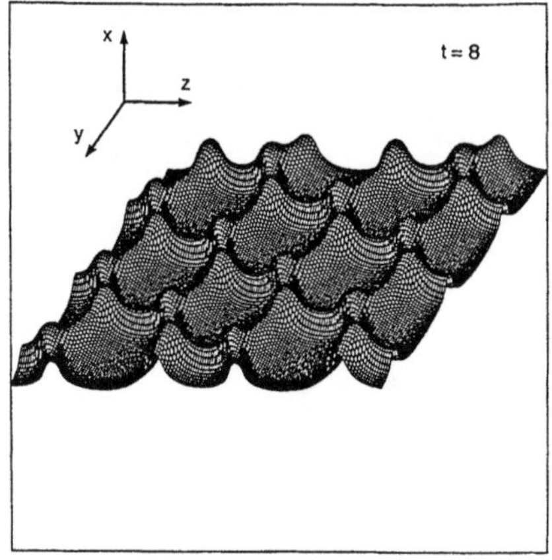

Fig. 7. Front shapes of the 3-D cellular flame for Le = 0.5, G = 0.0, λ_y = 13.3, and λ_z = 23.0 (t = 3 and 8); the schematic domain is $3\lambda_y\times2\lambda_z$.

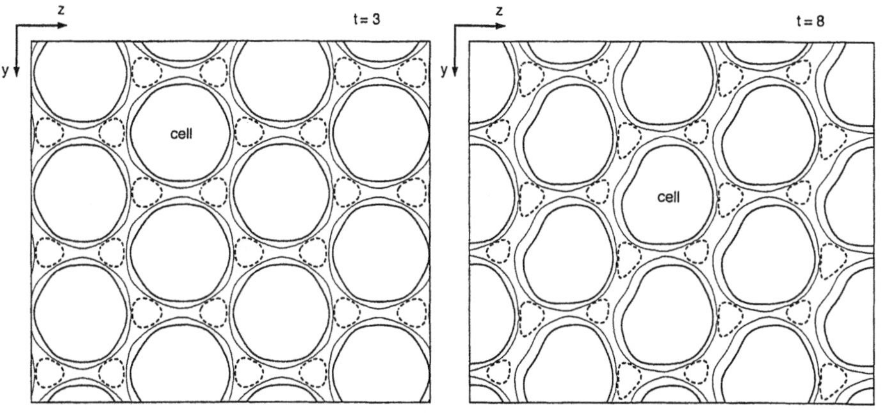

Fig. 8. Cell distributions of the 3-D cellular flame for Le = 0.5, G = 0.0, λ_y = 13.3, and λ_z = 23.0 (t = 3 and 8); the schematic domain is $3\lambda_y\times2\lambda_z$.

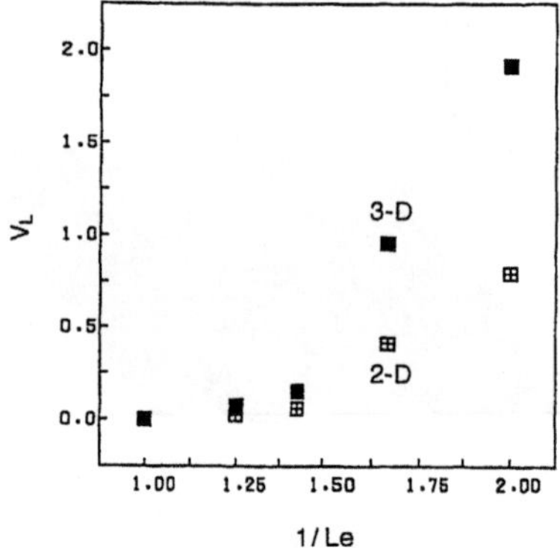

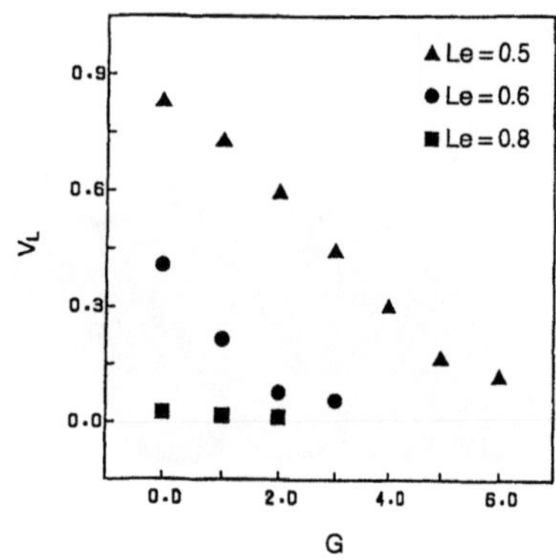

Fig. 9. Lateral velocities of 2-D and 3-D cellular flames, depending on the Lewis number.

Fig. 10. Lateral velocities of 2-D cellular flames for Le = 0.5, 0.6, and 0.8, depending on the acceleration.

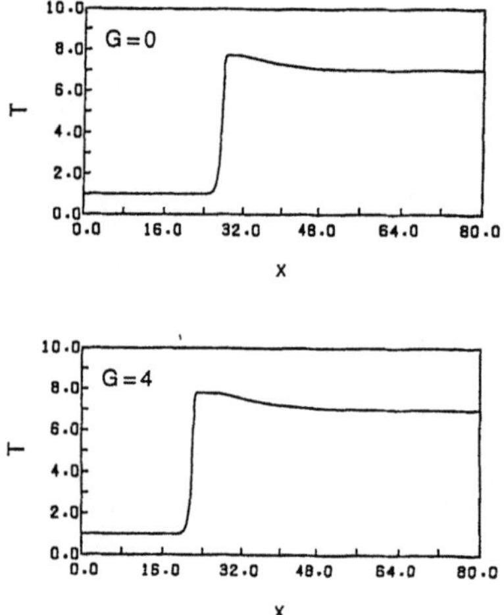

Fig. 11. Temperature distributions of 2-D cellular flames for Le = 0.5 and G = 0 and 4 at a convex flame front with respect to the unburned gas (t = 20).

Numerical Simulation of Combustion Processes in Homogeneous and Stratified Charge Spark Ignition Engines

Hiroshi Miyagawa, Yoshihiro Nomura and Makoto Koike

Toyata Central Research and Development Labs. Inc., Nagakute, Aichi, 480-1192, Japan

Summary. A three-dimensional simulation technique for stratified combustion process in direct injection gasoline engines is developed. The laminar flame speed for wide range of mixture equivalence ratio and EGR condition is modeled taking into account the reference temperature intermediate between unburned and flame temperature for chemical reaction. This new laminar flame speed model and the coherent flame model are incorporated into a CFD code. The calculated flame propagation process, heat release rate and exhaust emissions are validated by measurements including LIF technique. The good agreement obtained for various conditions shows the availability of this method.

 Key words. CFD, Stratified Charge, Direct Injection Gasoline Engine, Combustion, Laminar Flame Speed

Introduction

Recently, research and development of direct injection gasoline engines are actively carried out. In such engines, the fuel is injected directly into the cylinder at the intake stroke in order to prepare homogeneous mixture at relatively high speed and high load operation. On the other hand, under low speed and low load, the fuel is injected at the compression stroke to prepare stratified mixture. At stratified operation, it is necessary for stable combustion to make appropriate mixture around the ignition plug. This is a challenging task because the spray characteristics, chamber geometry and in-cylinder gas flow greatly affect the mixture formation process and after all they affect the engine performance and exhaust emissions. Therefore, it is expected to apply three-dimensional numerical calculation to the analysis of in-cylinder phenomena.

The CFD simulation of spray and mixture formation process has been practically used for the development of combustion system (Nomura et al. 1998).

In the next place, CFD simulation of combustion process is required to clarify the relation between the mixture distribution and engine performance.

In the last decade, several turbulent flame propagation models based on the flamelet concept have been proposed. In general, the fast chemical hypothesis is thought to be adequate for the combustion in spark ignition engines because the mixture is under relatively weak turbulence, high temperature and high pressure condition. Therefore, three dimensional numerical simulations using flamelet model are widely being validated for premixed engine combustion process (for instance Delhaya and Duverger 1998). We also showed the applicability of such a simulation method for development of combustion chamber geometry (Miyagawa et al. 1998).

In this study, previous method is improved for stratified combustion process in direct injection gasoline engines. In case of stratified charge combustion, the mixture is extensively distributed from rich to lean. The effect of mixture equivalence ratio on combustion is expressed via laminar flame speed in the flamelet model, so that a laminar flame speed model applicable for wide range of mixture equivalence ratio is developed. The applicability of the improved method for the prediction of combustion process and exhaust emission is investigated.

Flamelet model

Figure 1 shows an example of flame images in stratified combustion process using a bottom view type optical accessible engine. The time interval between fuel injection and ignition is fairly longer than that of diesel engines, so that the mixture is partially premixed. It is recognized that the principal combustion process is flame propagation rather than diffusion flame even in stratified operation of direct injection gasoline engine.

Therefore, flamelet model is employed as a combustion model as in the case with flame propagation in homogeneous mixture. Figure 2 is a schematic of turbulent propagating flame structure. The flame surface locally propagating at laminar flame speed is bent and stretched by turbulent eddies. For such flames, flamelet model describes the local fuel consumption rate R_f as follows:

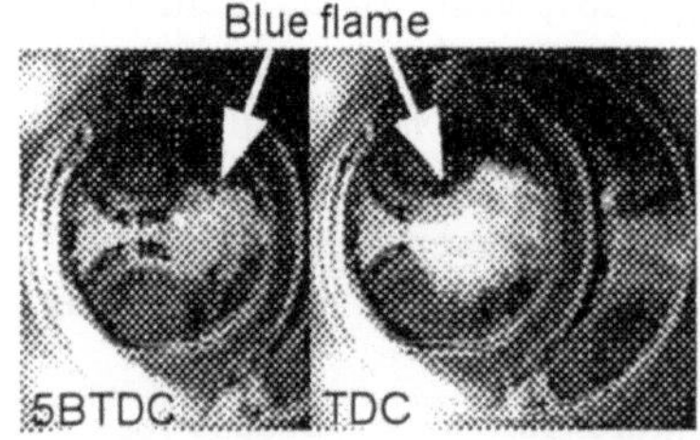

Fig. 1 Combustion under stratified operation

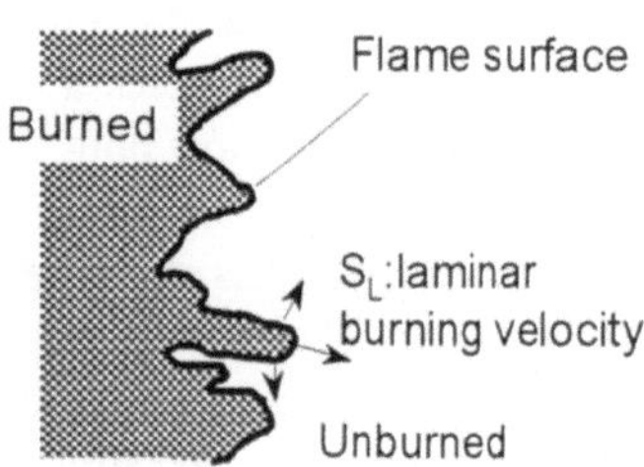

Fig. 2 Schematic of flame structure

$$R_f = \rho_{fu} S_L \Sigma \tag{1}$$

where ρ_{fu}, S_L and Σ are unburned fuel density, laminar flame speed and flame surface density, respectively. In this study, CFM (Coherent Flame Model, Duclos et al. 1993) is employed and a transport equation for Σ is solved. Using ITNFS (Intermittent Turbulence Net Flame Stretch, Meneveau and Poinsot 1991) model, the turbulent flame stretch rate is correlated with the characteristic scale ratio of turbulence and laminar flame (L/δ_L and u'/S_L where L, δ_L and u' are turbulent integral length scale, laminar flame thickness and turbulent intensity).

Laminar flame speed model

Figure 3 shows a mixture distribution measured with LIF(Laser Induced Fluorescence) method in an engine at stratified operation. The mixture equivalence ratio distributes widely from rich ($\phi>2$) to lean ($\phi<0.5$). Besides that, conventional three-way catalysts dose not work at the stratified operation, so that large amount of EGR is added for NOx reduction.

In the flamelet model, the effects of mixture composition are considered through the laminar flame speed. Therefore, laminar flame speed model applicable for wide range of mixture equivalence ratio or EGR condition is required in order to calculate stratified combustion process in direct injection gasoline engines. The widely used laminar flame speed models (for example Lavoie 1978, Metghalchi and Keck 1982 and Muller et al. 1997), however, are inadequate because their applicable ranges are limited around the stoichiometry and their model constants have to be adjusted for each mixture composition.

Tabaczynski et al. (1977) empirically modified a theoretical model (Van Tiggelen and Deckers 1957) based on a simplified chain branching scheme :

$$S_L = C\, p^{\alpha} T_u \left[X_f^{\ z} X_{O_2}^{\ 1-z} T_m^{\ -1} \exp\left(- E / RT_m \right) \right]^{0.5} \tag{2}$$

$$T_m = T_u + 0.74\left(T_b - T_u\right) \tag{3}$$

where p is pressure, T_u, T_b are unburned mixture temperature and burned gas

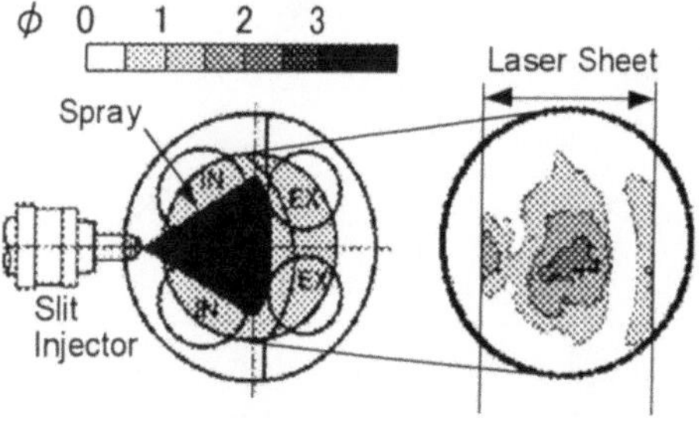

Fig. 3 Mixture distribution in a DI gasoline

temperature, X_f, X_{O2} are mole fraction of fuel and oxygen in unburned gas, respectively. C, α, Z, E are model constants correlated with measured data and depending on fuel type. These model constants are independent of mixture equivalence ratio. However, it is not clear whether this model is applicable for a richer and leaner mixture beyond the measured range and for a mixture with EGR.

In this study, a laminar flame speed model is developed as follows. It is derived from thermal theory that laminar flame speed is in proportion to the root of the mean reaction rate in the flame zone. Therefore, Arrhenius plots are used in order to investigate the relationship between measured laminar flame speed and reference temperature T_r for chemical reaction. T_r is defined by next equation.

$$T_r = T_u + c\left(T_a - T_u\right) \tag{4}$$

where T_a is adiabatic flame temperature. Figure 4 shows an example of Arrhenius plot for measured methane flame speed (Kurata et al. 1994). In case of c=1, that is T_r is equal to adiabatic flame temperature, the approximation lines for each mixture equivalence ratio are different. This means the model constants have to be adjusted for each mixture composition as in the most conventional models. On the other hand, all data are arranged with a single line when the constant c is chosen properly. This means the laminar flame speed is approximately expressed using simply unburned and burned gas temperatures for various mixture compositions. A laminar flame speed model is obtained as follows.

$$S_L = A\,p^\alpha \exp\left(-E/RT_r\right) \tag{5}$$

where A, α, E are model constants that do not need adjustment for mixture equivalence ratio or EGR amount. The model constant c in eq. (4) is a value between 0.33 and 0.5 depending on the fuel. It is interesting that the reference temperature T_r is close to the temperature at the inner edge of luminous zone and fairly lower than flame temperature.

Figure 5 shows a validation result of this model. The experimental result is due to Kurata et al. (1994) for methane flame. The model constants are obtained using the measured data for $0.8<\phi<1$. However, the model agrees well with the measurement for relatively wide range of equivalence ratio including $\phi<0.8$. It is confirmed that the laminar flame speed can be extrapolated using this model.

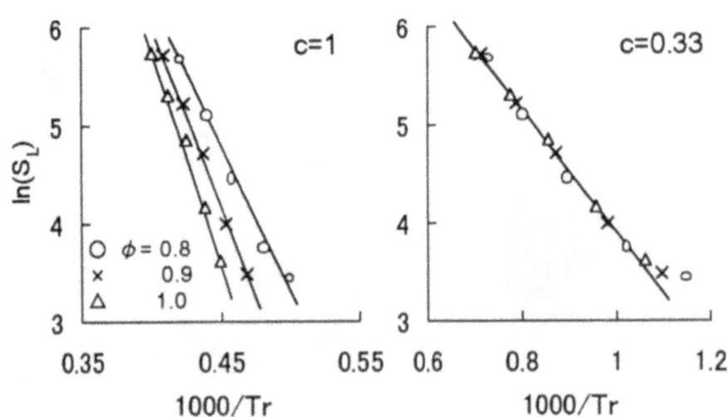

Fig. 4 Arrangement of measured laminar flame speed

The model constants are also obtained for iso-octane flame under high temperature and high pressure condition (Metghalchi and Keck 1982) in order to apply this model to engine combustion simulation. The present model is compared with the conventional models in figure 6. This model agrees well with the conventional models around stoichiometry while the model by Tabaczynski et al. (1977) overestimates the laminar flame speed. Figure 7 shows the model validation for EGR. This model predicts the EGR effect as well.

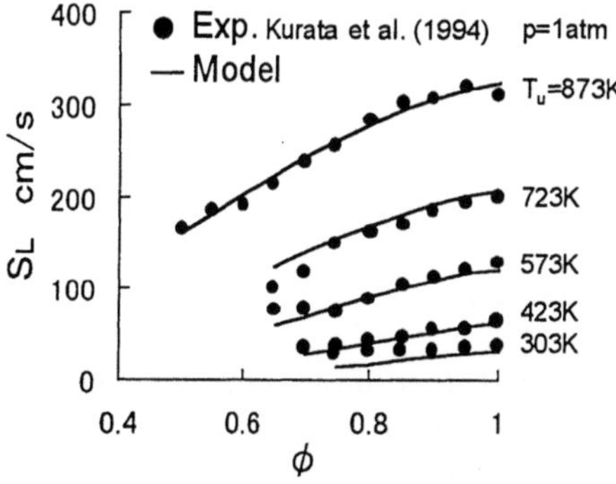

Fig. 5 Validation of laminar flame speed model

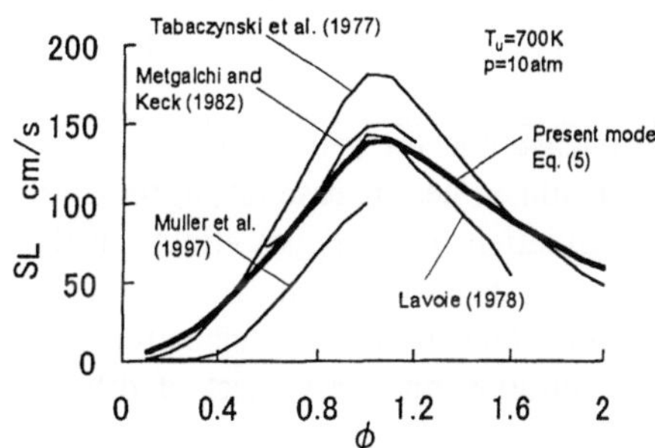

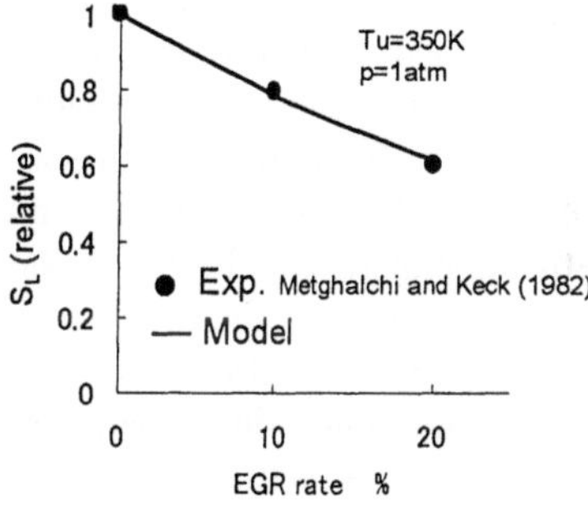

Fig. 6 Comparison of flame speed model

Fig. 7 Validation of laminar flame speed model

Calculation method

The basic equations are ensemble averaged three-dimensional Navier-Stokes equations, the transport equations of the standard k-ε model and the conservation equations of fuel mass and enthalpy. They are solved using a commercial STAR-CD code. The newly developed laminar flame speed model is incorporated into STAR-CD with CFM.

A flame kernel is assumed at the beginning of combustion. Both unburned and burned region are considered in each computational cell. Chemical equilibrium in burned gas is also assumed. NO emissions are calculated with the extended Zeldovitch mechanism and the reference temperature is not the cell averaged temperature but the burned gas temperature.

In case of stratified combustion, post flame reaction is not negligible. There are partial combustion species in rich burned gas while there is oxygen in lean burned gas. The post flame reaction occurs by mixing such gas cluster. In this calculation, the re-burning of rich burned gas is modeled with the change of equilibrium at mixing with lean burned gas or fresh air.

Results and discussions

Homogeneous combustion

First, the calculation method is validated for homogeneous combustion in premixed spark ignition engine. The calculations are compared with the experiments for various mixture equivalence ratio and EGR condition. In experiments, iso-octane is injected at the intake port. The calculation is carried out from the intake valve close timing (-160deg.ATDC) to 60deg.ATDC. Homogeneous mixture and solid vortex corresponding to the steady swirl ratio 2.2 are assumed as initial conditions. The engine specifications and calculation conditions are summarized in table 1.

Figure 8 shows the comparison of heat release rate for ϕ=0.7~1.2. The leaner mixture is, the slower combustion is. The calculated heat release rate agrees well with the experiment. Figure 9 shows the validation for inert gas effect. In this case, a part of air in the mixture is displaced with the nitrogen under constant G/F condition. Such a dilution simulates EGR condition in engine combustion. The heat release comes down with rising up the dilution rate. The effect of dilution with inert gas is relatively well calculated.

Table 1 PFI engine specifications and calculation conditions

Bore, Stroke	86mm, 86mm
Compression Ratio	9.7
Engine Speed	1200rpm
Volumetric Efficiency	40%
Ignition Timing	-30° ATDC

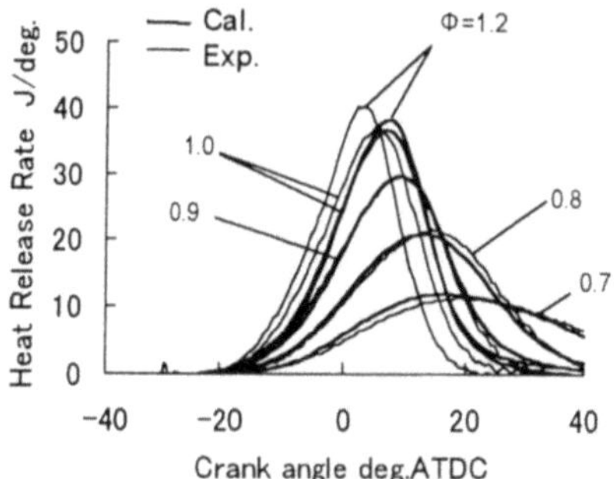

Fig. 8 Heat release rate for various equivalence ratio

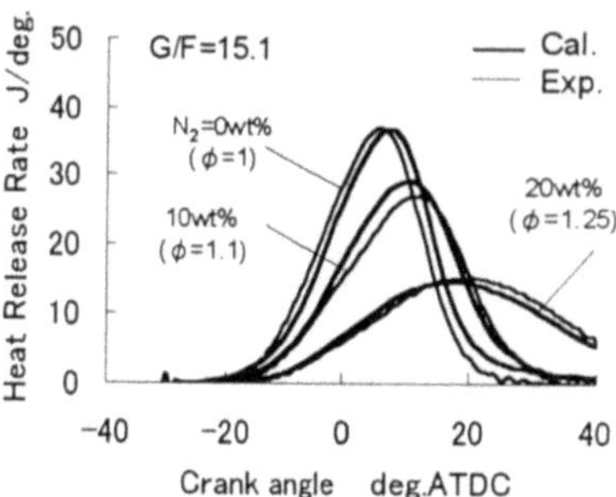

Fig. 9 Heat release rate for various dilution rate

Stratified combustion

The calculation is applied for stratified combustion in direct injection gasoline engine. The schematic of combustion system is shown in figure 10. A fan-shaped fuel spray is injected toward a shell-shaped piston cavity at the compression stroke. The mixture is guided to the cavity center using the cavity side wall. Finally, the stratified mixture is prepared around the ignition plug. The calculations of the mixture formation and combustion process are performed for prototype engines with such concept (Koike et al. 2000, Kanda et al. 2000). The discrete droplet model is employed for spray calculation (Nomura et al. 1998). The engine specifications and calculation conditions are summarized in table 2. In the calculation iso-octane is used as a substitute for gasoline used in the experiment.

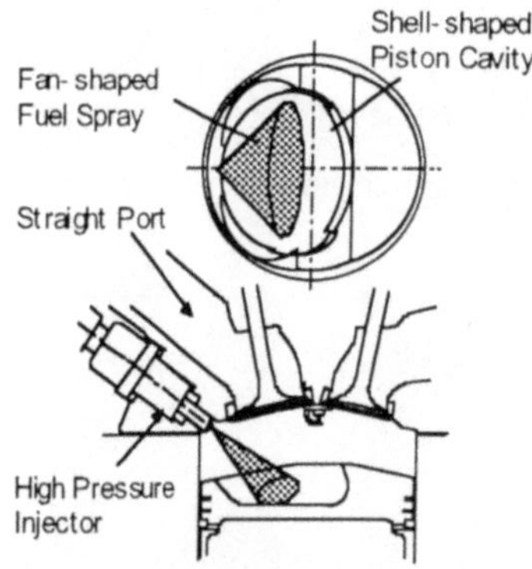

Fig. 10 Combustion system

Table 2 DI gasoline engine specifications and calculation conditions

Bore, Stroke	86mm, 86mm
Compression Ratio	10.7, 13.4
Engine Speed	1200, 2400rpm
Injection Pressure	12, 20MPa
Injection Timing	-80~-50° ATDC
Ignition Timing	-30, -12° ATDC

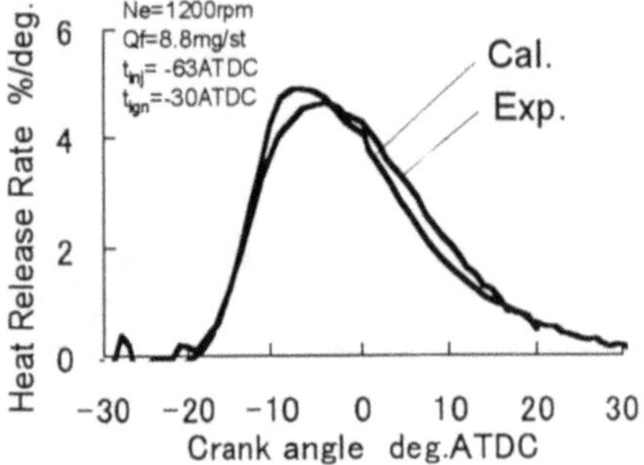

Fig. 11 Comparison of heat release rate

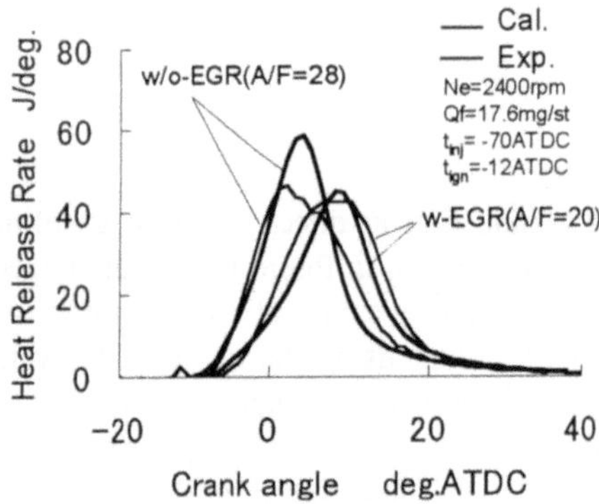

Fig. 12 Comparison of heat release rate

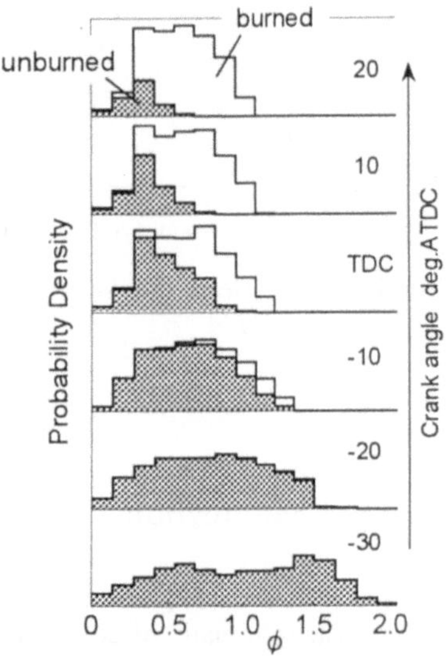

Fig. 14 Fuel distribution and consumption process

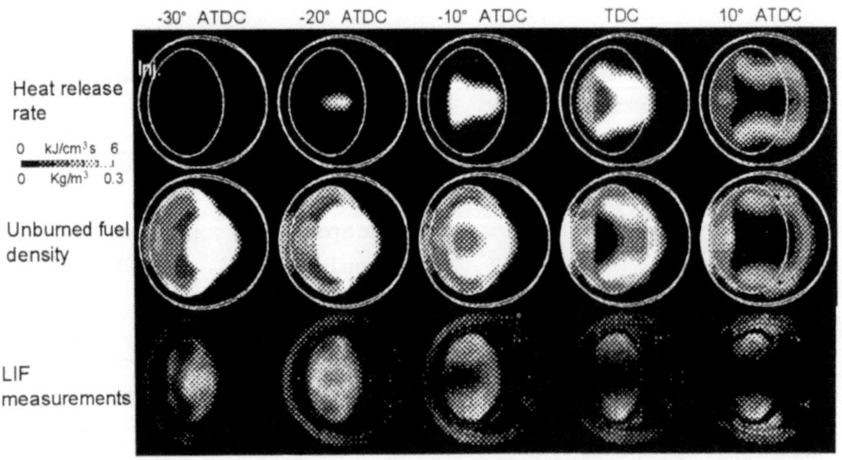

Fig. 13 Comparison of combustion process

Figure 11 shows the comparison of heat release rate at low speed and low load. After ignition delay period, heat release rises sharply at initial combustion phase while the combustion is relatively slow at later phase. Such heat release pattern is typical for low speed and low load in stratified operation. This characteristic is simulated fairly well. Figure 12 shows a validation result for EGR effect at middle speed and middle load. The EGR amount is approximately 30% of the mixture under with EGR condition. The combustion becomes slow with EGR. The effect of EGR is qualitatively calculated.

Figure 13 shows the comparison of unburned mixture distribution at combustion process. The calculated distribution of local heat release rate is also shown. The calculation condition is same as in figure 11. In the experiment, unburned mixture distribution is observed using a bottom view type optical accessible engine and LIF method (Fujikawa et al. 2000). The whole space of combustion chamber is illuminated by expanded laser beam. The absolute value of fluorescence intensity at each crank angle is different. In case of calculation, the distribution is expressed by the maximum fraction on the projected direction. At the first, the flame propagates toward the injector side (left side in figure) from the ignition point. The later, the flame expands to cavity side (up and down in figure). With such a flame propagation process, the fuel starts to be consumed from cavity center and unburned fuel remains at the cavity side. This process is consistent with the measurement.

Figure 14 shows the time history of fuel mass probability density against the equivalence ratio. The mixture concentration is distributed widely at −30 deg.ATDC. It is homogenized gradually with time. After TDC, though, the distribution shape changes little. The mixing becomes slower with the rise of burned mass fraction. The combustion takes place from relatively rich mixture because the mixture is stratified around the ignition point. In the later phase, over lean mixture remains and the combustion becomes slow. Such an over lean mixture is thought to be a source of unburned hydrocarbon.

Next, the availability of this simulation technique for prediction of exhaust emissions is discussed. In lean mixture, the combustion is not complete even after the end of main combustion as described above. In such a flame, quenching occurs at following expansion stroke because the turbulence becomes strong relative to the flame speed falling rapidly with the temperature decrease. In this calculation, flame quenching is described using ITNFS model (Meneveau and Poinsot 1991). In figure 15, the calculated unburned fuel amount at 40deg.ATDC is compared with measured engine-out THC for two types of piston cavity geometry. Cavity-II that is modified from cavity-I to control the over mixing, reduces HC emission. Both results are consistent. Figure 16 shows the comparison between calculated NO and measured one. The calculation is performed for fuel injection timing and injection pressure. The trend of NO production is predicted for various injection conditions.

Finally, the further model improvement concerning about the flame structure is discussed. Figure 17 shows a classification of the flame structure in the stratified combustion calculation. Each dot corresponding to computational cell in flame zone is plotted on a combustion diagram depending on the characteristic scales of turbulence and local laminar flame. Most of the flame belongs in classical flamelet regime, and also belongs in conventional engine flame regime (Abraham et al. 1985). The gray dots denote lean flames of which mixture equivalence ratio are under 0.6. In some of these lean flames, the Karlovitz number is over unity. The local flame speed in such flame is thought to be different from the laminar flame speed because turbulence affects heat and mass transfer in the preheat zone of such flame. This concerns the accuracy of unburned hydrocarbon attributable to lean mixture combustion and further study is required.

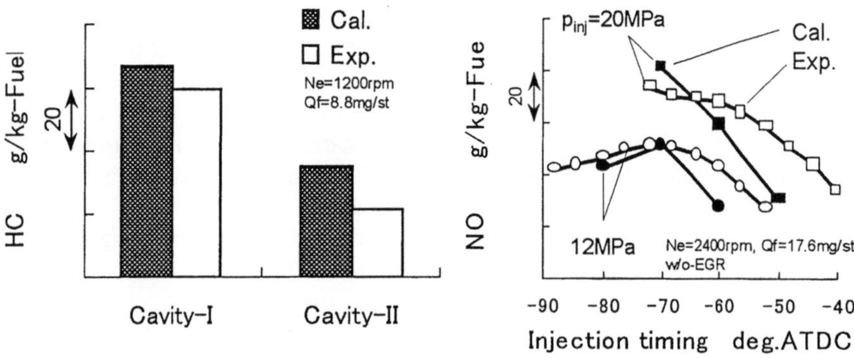

Fig. 15 Validation for HC emission   Fig. 16 Validation for NO emission

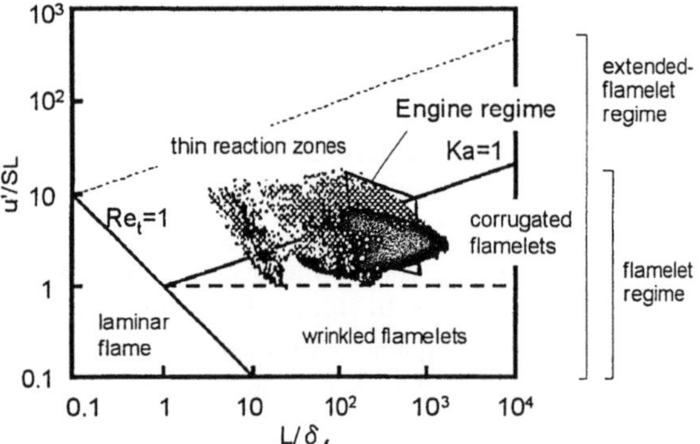

Fig. 17 Flame regime (4deg.ATDC, Ne=2400rom, Qf=17.6mg/st, w/o-EGR)

Conclusions

A three-dimensional simulation technique for stratified combustion process in direct injection gasoline engines is studied. The results are summarized as follows:

(1) The laminar burning velocity for wide range of mixture equivalence ratio and EGR condition is modeled taking into account the intermediate temperature between unburned and flame temperature as reference temperature for chemical reaction.

(2) Using present laminar flame speed model with flamelet model, stratified combustion process is simulated. The calculated heat release rate, unburned mixture distribution and exhaust emissions agree well with the experimental results for various conditions and specifications.

References

Abraham J, Williams F A, Bracco F V (1985) A discussion of turbulent flame structure in premixed charges. SAE paper 850345

Delhaye B, Duverger T (1998) Modeling of internal flow and combustion in a four valve lean-burn SI engine. The fourth international symposium COMODIA 98:261-267

Duclos J M, Veynante D (1993) A comparison of flamelet models for premixed turbulent combustion. Combustion and Flame 95:101-117

Fujikawa T, Hattori Y, Akihama K, Koike M (1998) Quantitative 2-D fuel distribution measurements in a direct-injection gasoline engine using laser-induced fluorescence technique. The fourth international symposium COMODIA 98:317-322

Fujikawa T, Nomura Y, Hattori Y, Kobayashi T, Kanda M (2000) Analysis of combustion fluctuation in a direct-injection gasoline engine using laser-induced fluorescence technique. The 16th Internal Combustion Engine Symosium:13-18 (in Japanese)

Kanda M, Baika T, Kato S, Iwamuro M, Koike M, Saito A (2000) Application of a new combustion concept to direct injection gasoline engine. SAE paper 2000-01-0531

Koike M, Saito A, Tomoda T, Yamamoto Y (2000) Research and development of a new direct injection gasoline engine. SAE paper 2000-01-0530

Kurata O, Takahashi S, Uchiyama Y (1994) Influence of preheat temperature on the laminar burning velocity of methane-air mixtures. SAE paper 942037

Lavoie G A (1978) Correlations of combustion data for S.I. engine calculations – Laminar flame speed, quench distance and global reaction rate. SAE paper 780229

Nomura Y, Fujikawa T, Koike M, Kanda M, Kobayashi T (1998) Prediction of mixture formation in the cylinder of DI gasoline engine. 76th JSME Fall Annual Meeting No.98-3:513 (in Japanese)

Meneveau C, Poinsot T (1991) Stretching and quenching of flamelets in premixed turbulent combustion. Combustion and Flame 86:311-332

Metghalchi M, Keck J C (1982) Burning velocities of mixtures of air with methanol, isooctane, and indolene at high pressure and temperature. Combustion and Flame 48:191-210

Miyagawa H, Kojima S, Katsumi N, Ueda T, Okumura T (1998) Numerical analysis of the effects of squish geometry on a newly developed 4-valve gasoline engine combustion process. The fourth international symposium COMODIA 98:227-232

Muller U C, Bollig M, Peters N (1997) Approximations for burning velocities and Markstein numbers for lean hydrocarbon and methanol flames. Combustion and Flame 108:349-356

Tabaczynski R J, ferguson C R (1977) A turbulent entrainment model for spark-ignition engine combustion. SAE paper 770647

Van Tiggelen A, Deckers J (1957) Chain branching and flame propagation. 6th Symp (Int) Combustion:61-66

Numerical Simulation of Hydrogen/Air Jet Diffusion Flame at NAL

Yasuhiro Mizobuchi, Ryoji Takaki and Satoru Ogawa

Computational Science Division, National Aerospace Laboratory, 7-44-1 Jindaiji-higashi, Chofu, Tokyo 182-8522, Japan

Summary. Diffusion flames by subsonic hydrogen jet injection into still air are numerically simulated with a fine grid system whose grid spacing is several times as large as the Kolmogorov scale. The result shows fairly good agreement with the experiment in the flame lift-off height and the calculated energy spectra are fairly sound. The fractal dimension of hydrogen mole fraction iso-surface is analyzed and the results show that the fractal dimension of the interface is 2.3-2.4 in unburned regions and decreases in burned regions, which coincides with the experimental measurements. The swirled injection effect is investigated and it is shown that the flame lift-off height becomes shorter by the swirl effect.

Key words. Numerical Simulation, Diffusion Jet Flame, Turbulence, Fractal Dimension

Introduction

Turbulent combustion is a very important phenomenon for human life and frequently observed in many engineering applications. From the viewpoint of future energy and environment issues, improvement of combustion efficiency and reduction of pollutant emission are indispensable. In Japan from FY 1999, three national research institutes, MEL (Mechanical Engineering Laboratory), SRI (Ship Research Institute) and NAL (National Aerospace Laboratory) have been conducting a research project on turbulence control in collaboration with universities and companies under the Open and Integrated Research Program by Science and Technology Agency of Japan. In this project, turbulent combustion control is one of the most important research topics and a working group is organized. The working group includes researchers of various research fields: fluid dynamics, combustion, control, measurement, computer and so on. Experimental, theoretical and numerical researches are conducted toward the establishment of clean and efficient combustion form from the viewpoints of fundamental study and applications.

To achieve highly clean and efficient combustion, we need a detailed understanding of the physics in turbulent combustion flow. However, turbulence phenomena, chemical reactions and their interaction are so complicated that investigation of turbulent combustion has been difficult. Recent drastic developments of computer technology enable us to conduct direct numerical simulations (DNS) of turbulent combustion flow and to get detailed data from the simulated flowfield. DNS can be an analysis tool of turbulent combustion.

At NAL, large-scale numerical simulations of jet diffusion flame have been conducted using the parallel-vector computer NWT (Numerical Wind Tunnel) with a very fine grid system of over 20 million grid points to investigate the phenomena in hydrogen/air system. In such a large computation, post-processing to derive information from the huge computed data is important. Statistical approach is an ordinary method for turbulence-related problems, but for this case it is difficult at present because it needs a time-series of computational data, which is too large even for today's computational resource. Now we focus on the fractal dimension of the flowfield, which can be defined from instantaneous data. Fractal is thought to be one property of turbulence and has been measured in turbulence experiments. Prasad and Sreenivasan (1990) measured the fractal dimensions of various types of turbulent flows. Yoshida et al. (1994) measured the fractal dimension of premixed flame front and analyzed its dependency on turbulence intensity. The fractal analysis of turbulent combustion flow is expected to bring some information about the turbulence effects on mixing and chemical reaction.

In this paper the results of the detailed numerical simulation of hydrogen/air jet diffusion flames are shown. The fractal dimension analysis is presented and nominal results of statistical post-processing, energy spectrum analysis, are also shown. As a preliminary step of control research, the effects of swirled injection are investigated.

Problem configuration

The configuration of the problem is shown in "**Fig. 1**". A subsonic hydrogen jet is injected into still air from a round nozzle. The configuration is subject to the experiment by Cheng et al. (1992). The nozzle diameter D is 2mm. The pressure and temperature of the jet are 1atm and 280K, respectively. The jet velocity v_{jet} is 680 m/sec, the Mach number is 0.54 and the Reynolds number based on the nozzle diameter is 13600. The temperature and pressure of the still air are the same as those of the jet. In the experiment the lifted flame is observed and the lift-off height is about 7 diameters.

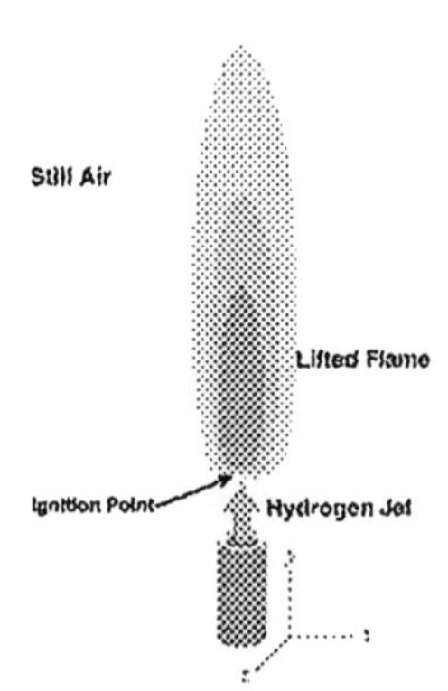

Fig. 1. Schematics of problem.

Computational model and governing equations

The flowfield is assumed to be in thermal equilibrium and in chemical nonequilibrium. The 9-species (H_2, O_2, OH, H_2O, H, O, H_2O_2, HO_2, N_2) and 17-reaction model by Westbrook (1982) is employed as a chemical reaction model. The transport coefficients of each chemical species are estimated using the Lennard-Jones intermolecular potential model and those of the gas mixture are calculated by Wilke's empirical law: Wilke(1950). The diffusion flux of chemical species is evaluated using Fick's law. The enthalpy of each chemical species is quoted from the JANAF thermochemical table (1965). Gravitational effects are ignored because the jet speed is high enough and no turbulence model is used.

The governing equations are written for a generalized curvilinear coordinate system as,

$$\frac{\partial Q}{\partial \tau} + \frac{\partial F_{\xi_i}}{\partial \xi_i} = \frac{\partial F_{v\xi_i}}{\partial \xi_i} + H_c \;,$$

where,

$$Q = V \begin{bmatrix} \rho \\ \rho u_1 \\ \rho u_2 \\ \rho u_3 \\ E \\ \rho z_s \end{bmatrix}, \quad F_{\xi_i} = \begin{bmatrix} n_{ij}\rho u_j \\ n_{ij}\rho u_1 u_j + n_{i1}p \\ n_{ij}\rho u_2 u_j + n_{i2}p \\ n_{ij}\rho u_3 u_j + n_{i3}p \\ n_{ij}(E+p)u_j \\ n_{ij}\rho z_s u_j \end{bmatrix}, \quad F_{v\xi_i} = \begin{bmatrix} 0 \\ n_{ij}\tau_{1j} \\ n_{ij}\tau_{2j} \\ n_{ij}\tau_{3j} \\ n_{ij}(\tau_{jk}u_k + q_j) \\ n_{ij}\rho D_s z_{s,j} \end{bmatrix},$$

$$E = e + \frac{1}{2}\rho \sum_{j=1}^{3} u_j^2 \;, \quad e = \sum_s \rho z_s \left(H_s + \Delta H_{fs}\right) - p \;, \quad p = R_u T \sum_s \rho z_s \;,$$

$$q_j = \kappa T_{,j} + \sum_s \rho D_s h_s z_{s,j} \;, \quad \tau_{ij} = \mu \left\{ u_{i,j} + u_{j,i} - \frac{2}{3}\delta_{ij}u_{m,m} \right\} \;,$$

where z_s, H_s and ΔH_{fs} are the mole number density per unit mass, the enthalpy and the heat of formation per mole of species s, respectively and u_i is the velocity component in the x_i direction. R_u is the universal gas constant. H_c is the source term by the chemical reactions. $(x_1, x_2, x_3) \equiv (x, y, z)$ and $(\xi_1, \xi_2, \xi_3) \equiv (\xi, \eta, \zeta)$ are Cartesian and curvilinear coordinate systems, respectively, $(\;\;)_{,i} \equiv \partial (\;\;)/\partial x_i$, V is the cell volume and n_{ij} is the cell-interface normal vector. Other notations without special mention are subject to common usage.

Computational method

Discretization method

The governing equations are discretized in a finite-volume-like formulation. The convective terms are evaluated using a TVD (Total Variation Diminishing) numerical flux which is based on Roe's approximate Riemann solver: Roe (1981) and developed for chemically reacting gas by Wada et al. (1989). The flux should be anti-dissipative so as not to damp small disturbances while it should be robust. Here, the higher-order flux is constructed extrapolating the characteristics with flux limiters proposed by Wada (1994). The numerical flux on a cell-interface $F_{j+1/2}$ is written as follows,

$$F_{j+1/2} = \frac{1}{2}\left[F_j + F_{j+1} - P_{j+1/2}\left(\Lambda^+_{j+1/2}\sigma^+_{j+1/2} - \Lambda^-_{j+1/2}\sigma^-_{j+1/2}\right)\right] ,$$

where,

$$\sigma^+_{j+1/2} = \sigma_{j+1/2} - \Psi(R_p)\sigma_{j-1/2} - \frac{2}{3}\overline{\Psi}(R_p)(\sigma_{j+1/2} - \sigma_{j-1/2}),$$

$$\sigma^-_{j+1/2} = \sigma_{j+1/2} - \Psi(R_m)\sigma_{j+3/2} - \frac{2}{3}\overline{\Psi}(R_m)(\sigma_{j+1/2} - \sigma_{j+3/2}),$$

$$\Psi(R) = \min\left\{-\frac{2\beta_2}{\min(R,-\varepsilon)}, 1, \frac{6\beta_1}{\max(R+2,\varepsilon)}\right\},$$

$$\overline{\Psi}(R) = \min\left\{\frac{3\max(0,R)}{\max(1-R,\varepsilon)}(\alpha-0.5), 1, \frac{6\beta_1}{\max(R+2,\varepsilon)}\right\},$$

$$\sigma = P^{-1}\Delta q , \quad R_p = \sigma_{j-1/2}/\sigma_{j+1/2} , \quad R_m = \sigma_{j+3/2}/\sigma_{j+1/2} ,$$

where, $q \equiv Q/V$, P and Λ are the eigenvector and eigenvalue matrices of $\partial F/\partial q$, respectively. The parameters α, β_1 and β_2 are chosen to be 1.0, 1.0 and 0.5, respectively and ε is a small positive value. The accuracy of this flux is third-order in smooth regions and keeps second-order even in regions where the sign of characteristics gradient changes. But it falls into first-order if the gradient change is extreme.

The viscous terms are evaluated with standard second-order difference formulae. The evaluation of diffusion terms should be a bit different from that of viscous terms. Diffusion fluxes are calculated with Fick's law, which is, exactly speaking, correct only when the mixture is binary. When it is applied to a mixture of more than three components, the sum of diffusion fluxes of all chemical species can not be zero unless all diffusion coefficients are the same, although it should be zero to conserve the total mass. Here the diffusion fluxes are modified so that the sum of positive fluxes and that of negative have the same absolute value as,

$$\hat{f}_s^{\,dif} = f_s'^{+} - f_s'^{-} ,$$

$$f_s'^{\,\pm} = f_s^{\,\pm} \cdot S^{ave}/S^{\pm} , \quad S^{ave} = (S^{+} + S^{-})/2 ,$$

$$S^{\pm} = \sum_s f_s^{\,\pm} , \quad f_s^{\,\pm} = (\, |f_s^{\,dif}| \pm f_s^{\,dif} \,) / 2 ,$$

where, $f_s^{\,dif}$ and $\hat{f}_s^{\,dif}$ are Fick's diffusion flux and the modified flux of species s, respectively.

The time-integration method is the explicit Runge-Kutta multi-stage method. The space-discretization accuracy is at most third-order and second-order in the vicinity of vortices. In wave equations, the truncation errors due to space-discretization and due to time-discretization are equivalent and therefore it is not so effective to use higher than third-order time-integration. The accuracy of time-integration used here is second-order.

Boundary conditions

The nozzle tube is rectangular for computational simplicity although it is cylindrical in the experiment. The surfaces of the nozzle tube are assumed to be slip walls and on the top of the tube the jet exit condition is imposed in $r \equiv (x^2 + z^2)^{1/2} < D/2$. On the jet exit the y-direction velocity is extrapolated. The total pressure and the total temperature are fixed so that the velocity profile has a 1/7 power law boundary layer in $3/8D < r < D/2$. No artificial disturbance is imposed at the jet exit boundary. At the outer boundaries the non-reflection condition by Thompson is applied so that the pressure waves successfully pass through the boundaries. At the beginning of computation, the entire computational region is filled with still air and after the flowfield is established, heat is added around $y=7D$ for ignition.

Computational grid system

The computational region is $-12D<x, z<12D, -3D<y<20D$, where the coordinate axes are subject to **"Fig. 1"** and the origin is the nozzle exit center. The grid system is rectangular and the grid point numbers are 205, 505 and 205 in x, y and z directions, respectively. The grid spacing is 0.05mm in $-1.25D<x, z<1.25D$, $0<y<8D$. This size is 2.5times as large as the Kolmogorov scale around the ignition point measured in the experiment and the Kolmogorov scale becomes smaller as the distance from the jet exit is shorter. The grid size of this computation is hence several times larger than the Kolmogorov scale and therefore, the smallest structure cannot be resolved. But recently it has been reported by Moin and Mahesh (1998) that the grid size can be several times as large as the Kolmogorov scale to simulate turbulent flows directly. In this computation the grid size is determined optimistically based on such research results. The grid spacing is coarser as the distance from the above mentioned region is longer.

The computation is conducted using 48 processor elements of NWT. The computational region is decomposed into 48 blocks and one processor element is assigned to one block.

Results and discussion

Over view of flowfield

"Fig. 2" is the iso-surface of hydrogen mole fraction at 70% at some moment. The flowfield is strongly three-dimensional and hydrogen/air mixing has a very complicated form. The iso-surface of temperature at 1000K is shown in **"Fig. 3 "**. A lifted flame is demonstrated and the lift-off height is 5.0 - 5.5 diameters while it is about 7 diameters in the experiment. This agreement is fairly good considering the difficulty of the problem. It is achieved by the resolution of such a small structure as several times of the Kolmogorov scale. The hydrogen and air are mixed sufficiently to maintain the combustion reactions and the momentum is also mixed well so that the velocity in the reacting region is small enough to hold a stationary lifted flame.

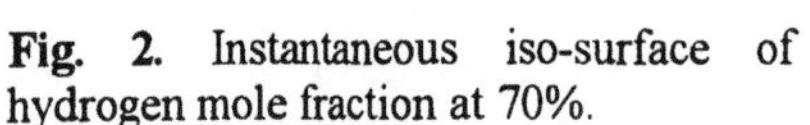

Fig. 2. Instantaneous iso-surface of hydrogen mole fraction at 70%.

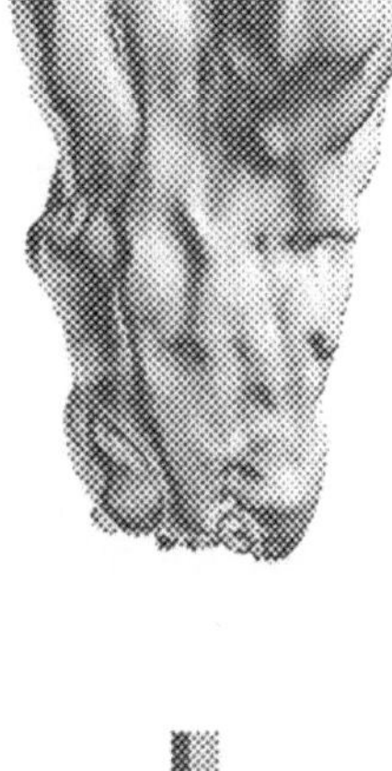

Fig. 3. Instantaneous iso-surface of temperature at 1000 K.

The contour lines of temperature from 400 to 2600 K in a cross section that includes the nozzle center are shown in **"Fig. 4"**. The shape of the flame outer side is dull as observed in **"Fig. 3"**, but the inner side has a very complicated structure and the complexity decreases as the distance from the jet exit increases, that is, as the combustion reactions proceed. The vorticity magnitude iso-surface is shown in **"Fig. 5"**. The vorticity rapidly decreases in the burned region, which also indicates the decrease of complexity of the flow structure along with combustion.

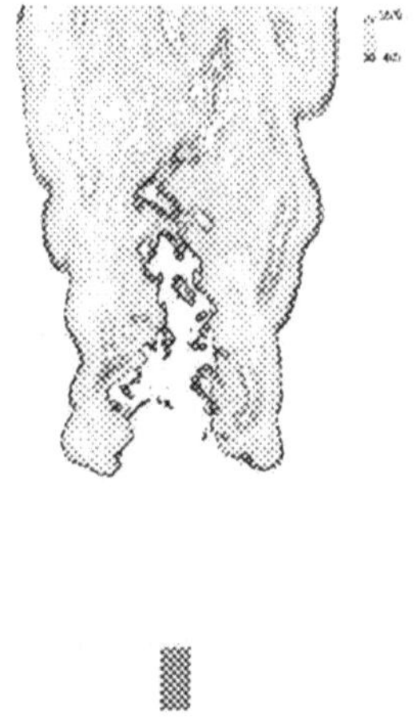

<table>
<tr><td>

Fig. 4. Instantaneous contour lines of temperature from 400 to 2600 K in a cross section that includes the nozzle exit center.

</td><td>

Fig. 5. Instantaneous iso-surface of vorticity magnitude at $0.4v_{jet}/D$.

</td></tr>
</table>

Energy spectrum

Statistical post-processing requires a long time series of three-dimensional data and enormous computational time and storage. To conduct statistical analysis, we need to divide instantaneous value into averaged value and fluctuation. But for essentially unsteady problems, it is difficult to define the average state because the average state can change with the sampling period.

Here, we tried the energy spectrum analysis about one spatial point, $x=y=0$, $y=4D$ for two sampling periods and rates. In the short period sampling case, sampling period is 0.03msec and the sampling rate is 2MHz, where the sampling interval is smaller than the measured Kolmogorov time scale and the sampling period is larger than the large-eddy-turnover time. In the long period case, the period is 0.3msec and the rate is 0.2MHz, where the sampling interval is larger than the Kolmogorov time scale and smaller than the large-eddy-turnover time.

"Figs. 6 and 7" show the energy spectra in three directions for the short and long period sampling cases, respectively. The energy spectra for the two cases are similar and the results are almost independent of the sampling period and rate. The results show that turbulence is homogeneous in the x-z plane and inhomogeneous in the y direction and that the computed energy spectra are fairly sound and most of turbulent energy is captured in computation.

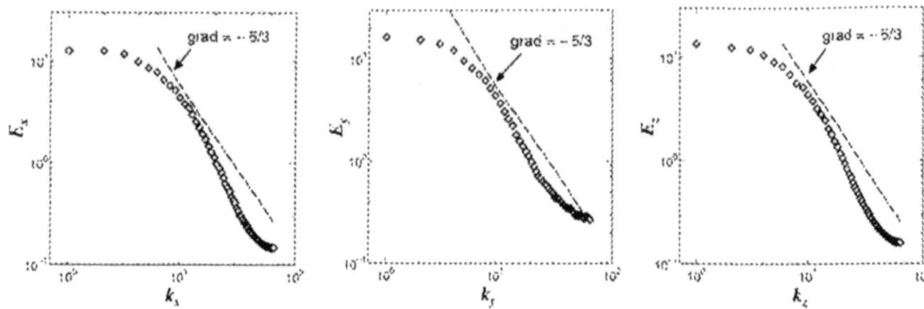

Fig.   6. Energy spectra at $x=z=0$, $y=4D$ measured over 0.03msec at 2MHz sampling rate.

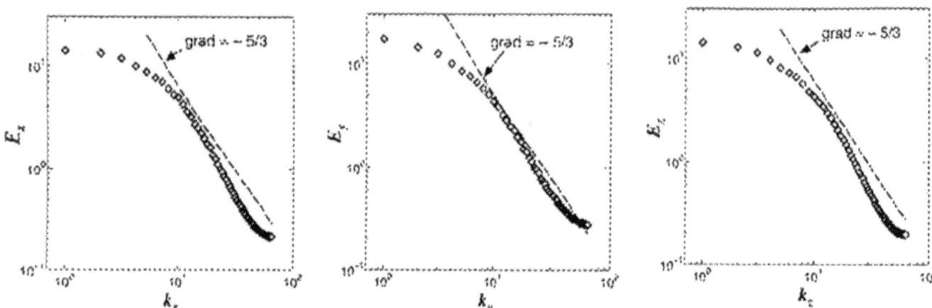

Fig.   7. Energy spectra at $x=z=0$,$y=4D$ measured over 0.3msec at 0.2MHz sampling rate.

Fractal dimension

Fractal is thought to be one property of turbulence. The fractal dimension D_f of an interface is defined as,

$$N \propto r^{-D_f} \, ,$$

where r is the size of the surface elements used in the measurement and N is the number of the elements needed to cover the interface. The fractal dimension is actively measured in turbulence experiments. It is experimentally reported that the fractal dimension of turbulence is 2.3-2.4 and 2.36 for round jet flows by Prasad Sreenivasan (1990).

"**Fig.   8**" shows the hydrogen mole fraction iso-surface at 80% in $2D < y < 8D$ at some moment. The OH distribution is also drawn, but no OH is observed on the surface. The fractal dimension of this surface can be calculated by measuring the surface area using various sizes of surface elements. The fractal dimension analysis of the surface is shown in "**Fig. 9**". The gradient of the fitted line should be the fractal dimension, and it is 2.36. The fractal dimension of the hydrogen mole fraction iso-surface varies with the iso-surface level and fluctuates with time, but this agreement with the measurement validates our computational technique. The plot at the smallest element size is below the fitted line. At present it is not sure whether it corresponds to the inner cut-off of fractal or it is due to numerical dissipation. More detailed investigation is needed.

Fig. 8. Instantaneous iso-surface of hydrogen mole-fraction at 80% with OH distribution in $2D<y<8D$.

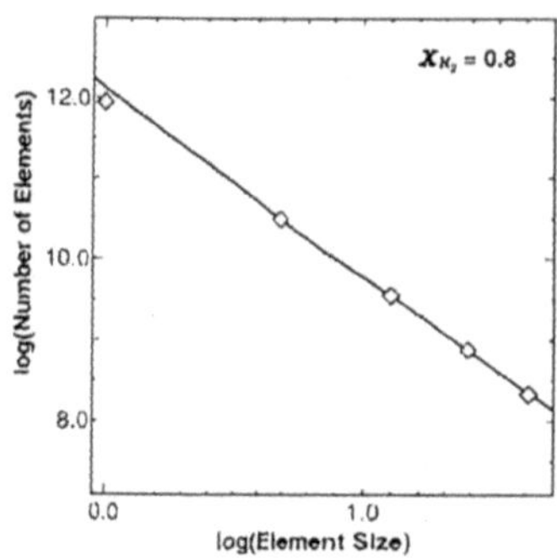

Fig. 9. Fractal dimension analysis of iso-surface of hydrogen mole-fraction at 80%, fractal dimension =2.36.

"**Fig. 10**" presents the hydrogen mole fraction iso-surface at 50% in $2D < y < 8D$ with OH distribution on the surface. OH is produced in $5D < y < 8D$ and it indicates that combustion reactions take place and the temperature is high in the region.

The fractal dimensions are calculated in two regions, namely the unburned region $(2D < y < 5D)$ and the burned region $(5D < y < 8D)$. They are 2.30 and 2.09 in the unburned region and in the burned region, respectively. On the other hand, the fractal dimension of the iso-surface at 80% is about 2.36 in both regions as listed in "**Table 1**". The fractal dimension of a burned surface is remarkably small compared with those of unburned surfaces. This result is strongly related to the complexity decrease with the proceeding of reactions as mentioned previously in "**Figs. 4 and 5**".

This tendency coincides with the experimental result by Yoshida et.al. (1994) that the fractal dimension of wrinkled laminar flames in turbulent premixed combustion flow is smaller than fractal dimension of non-reacting turbulent flows.

Molecular viscosity in the burned region increases due to temperature rise and, therefore viscous effects suppress turbulence and unevenness of interfaces. It is one reason for the fractal dimension decrease on burned surfaces. Further investigations into this result are needed and should be reflected on the modeling of turbulent combustion flows.

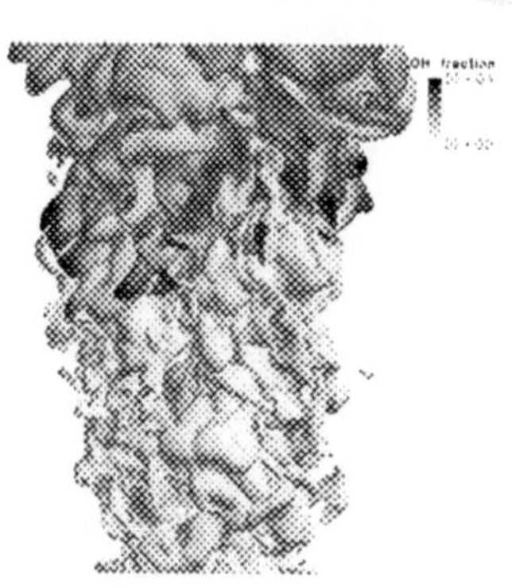

Fig. 10. Instantaneous iso-surface of hydrogen mole-fraction at 50% with OH distribution in $2D<y<8D$.

Table 1. Fractal dimensions of burned and unburned surfaces.

Iso-surface level of hydrogen mole fraction	Fractal dimension in $2D < y < 5D$	Fractal dimension in $2D < y < 5D$
0.5	2.30 (unburned)	2.09(burned)
0.8	2.36 (unburned)	2.36(unburned)

Effect of swirled injection

As a basic research of turbulent combustion control, the effects of swirled injection are numerically investigated. The angle velocity of injection jet is given so that the swirl velocity at the nozzle edge is 340m/sec.

The iso-surface of temperature at 1000K is presented in **"Fig. 11"**. The lift-off height is about $1D$ shorter than that in the normal injection case, which is shown in **"Fig. 3"**. The iso-surfaces of the second invariant of the velocity gradient at a positive value in $1D < y < 8D$ is shown in **"Fig. 12"**, **a**: normal injection case, **b**: swirled injection case. The positive second invariant corresponds to the rigid rotating region: Soria et al. (1994). The number of eddies becomes larger and the eddy structure more complicated by the swirled injection. Such effects enhance the hydrogen/air mixing and thus the lift-off height becomes shorter.

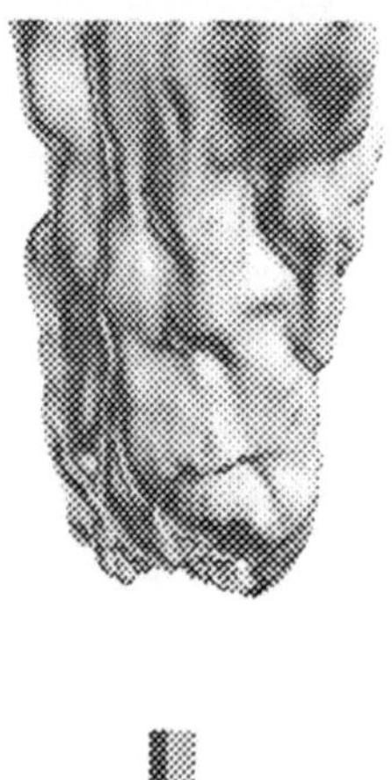

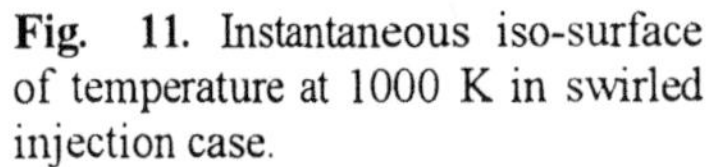

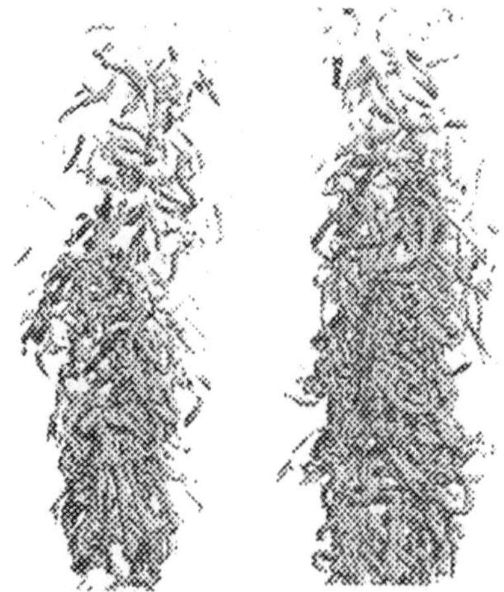

Fig. 11. Instantaneous iso-surface of temperature at 1000 K in swirled injection case.

Fig. 12a,b. Iso-surface of second invariant of velocity gradient tensor at a positive value in $1D<y<8D$. **a** normal injection case. **b** swirled injection case.

Concluding remarks

Diffusion flames by subsonic hydrogen jet injection into still air are numerically simulated with a fine grid system whose grid spacing is of the order of the Kolmogorov scale. Then the following major conclusions are obtained;
- The computed lift-off height fairly agrees with the measurement.
- The computed energy spectra are fairly sound.
- The computed fractal dimension of the jet flow interface is 2.3-2.4 and agrees with the measurement.
- The fractal dimension decreases on burned surfaces.
- The swirled injection results in a shorter flame lift-off height.

In future study, premixed lean combustion, especially combustion-driven oscillation, should be simulated. The control of such instability in turbulent combustion flow should be also simulated toward the construction of turbulent combustion control schemes.

Acknowledgments

This research has been carried out in the STA-funded research project "Smart Control of Turbulence: A Millennium Challenge for Innovative Thermal and Fluids Systems" at the Center for Smart Control of Turbulence.

References

Cheng T. S., Wehrmeyer J. A., Pitz R. W. (1992) Simultaneous temperature and multispecies measurement in a lifted hydrogen diffusion flame, Comb. and Flame, Vol.91, pp.323-345.

JANAF Thermochemical tables (1965).

Moin P. , Mahesh K. (1998) Direct numerical simulation : a tool in turbulence research, Annual Review of Fluid Mechanics, Vol.30, pp.539-578.

Prasad R. R., Sreenivasan K. R. (1990) The measurement and interpretation of fractal dimensions of the scalar interface in turbulent flows, Phys. Fluids A5, pp.792-807.

Roe P.L.(1981) Approximate Riemann solvers, parameter vectors, and difference scheme. J. Comp. Phys., Vol.43, pp.357-372.

Soria J., Sondergaard R., Cantwell B.J., Chong M.S., Perry A.E. (1994) Study of the fine-scale motions of incompressible time-developing mixing layers. Phys Fluids A6, pp.871-884.

Thompson K. W. (1987) Time dependent boundary conditions for hyperbolic systems,   J. Comp. Phys., Vol.68, pp.1-24.

Yoshida A., Ando Y., Yanagisawa T, Tsuji H. (1994) Fractal behaviour of wrinkled laminar flame, Combust. Sci. and Tech., Vol.96, pp.121-134.

Wada Y., Ogawa S. and Ishiguro T. (1989) A generalized Roe's approximate Riemann solver for chemically reacting flows. AIAA paper 89-0202.

Wada Y. (1995) Numerical simulation of high-temperature gas flows by diagonalization of gasdynamic matrices, Doctoral thesis, the university of Tokyo.

Westbrook C. K. (1982) Hydrogen oxidation kinetics in gaseous detonations. Combust. Sci. and Tech., Vol.29, pp.67-81.

Wilke, C. R. (1950) A viscosity equation for gas mixtures. J. Chem. Phys., Vol.18, No.4, pp.517-519.

Index

MIX
Papier aus verantwortungsvollen Quellen
Paper from responsible sources
FSC® C105338

If you have any concerns about our products,
you can contact us on
ProductSafety@springernature.com

In case Publisher is established outside the EU,
the EU authorized representative is:
**Springer Nature Customer Service Center GmbH
Europaplatz 3, 69115 Heidelberg, Germany**

Printed by Libri Plureos GmbH
in Hamburg, Germany